The INVENTOR'S HANDBOOK

SECOND EDITION

How to Develop, Protect, & Market Your Invention

Robert Park

BETTERWAY PUBLICATIONS, INC.
WHITE HALL, VIRGINIA

Published by Betterway Publications, Inc.
P.O. Box 219
Crozet, VA 22932
(804) 823-5661

Cover design by Susan Riley
Typography by Park Lane Associates

This book is designed to provide useful information and is sold with the understanding that neither the author nor the publisher is engaged in rendering legal, financial, or other professional services. If such expert assistance is required, the services of a competent professional person should be sought.

Library of Congress Cataloging-in-Publication Data

Park, Robert
     The inventor's handbook : how to develop, protect, and market your invention / Robert Park. -- 2nd ed.
          p. cm.
     ISBN 1-55870-149-4 : $12.95
     1. Inventions--Handbooks, manuals, etc. 2. Patents--Handbooks, manuals, etc. I. Title.
T339.P225 1990
608.773--dc20   89-18348
                                                             CIP

Printed in the United States of America
0 9 8 7 6 5 4 3 2

To "Corky" still,
for his continued patience and support,
and my son Robert Jr.,
for his stimulating company
these past two years.

Foreword

As one who has devoted much of his life to encouraging invention, helping inventors market their creations, and in general seeking ways to foster more innovation, it gives me great pleasure to write the foreword for *The Inventor's Handbook*.

It should be noted that innovation is the process of creating and utilizing technology to create new products, services and processes, or to make improvements in existing ones and bring them to market. Since the actual invention is a critically important and difficult part of the innovation process, *The Inventor's Handbook* can play a significant role in helping to facilitate invention and, in time, innovation. *The Inventor's Handbook* not only helps to successfully guide individuals along the complex path of invention, but also encourages them to market their inventions by forming a company for that purpose. That advice tracks with my own experience. I played a major role in starting two successful companies—one of them Control Data—and the experience was highly rewarding.

But as *The Inventor's Handbook* accurately notes, the transition from the laboratory to forming a successful company to market inventions is not an easy or simple one, as thousands of inventors and would-be entrepreneurs have learned. This book should provide invaluable guidance around some—but not all—of the obstacles to entrepreneurial success and self-sufficiency.

One of the most important results of the process of innovation is that everyone gains from inventions which are successfully marketed. In addition to the rewards to the inventor, society benefits even more from the jobs created and resulting increased economic activity. As Ralph Waldo Emerson said, "Invention breeds invention." So it's win, win for everybody.

Robert Park has written a valuable guide for inventors and has performed a great service to society. I am confident that *The Inventor's Handbook* will prove to be an important and lasting contribution to encouraging invention and facilitating the process of innovation.

William C. Norris,
Chairman Emeritus
Control Data Corporation

Contents

Preface to the Second Edition

Since *The Inventor's Handbook* was first published in the Fall of 1987, a great additional public awareness has grown over the importance of invention and innovation to the strength of our economy and our nation's vitality. For one thing, it has become more and more apparent that foreign technology is highly competitive to our own, and in ways which are drastically changing the very structure of our system of enterprise and our culture as a whole. We see it in the automobiles traveling our highways, in the cameras we take pictures with, in the radios and TVs we listen to and watch, in the clothes we wear, in the toys and sporting goods we enjoy and in every walk of life. Virtually all these products are manufactured now by overseas companies outcompeting our own. We see it too in the decline of patent applications being filed by Americans and in the concurrent increase in those filed by foreign nationals with the United States Patent and Trademark Office. All of this is alarming enough, but there's more:

Manufacturing has declined in America, and with it, so have manufacturing jobs in basic industries of all kinds. By consequence, many factories have closed and in rural America, entire communities have become devastated with serious income losses to their economies. I think this is shameful!

I have an abiding belief that the greatest of all our resources is our creativity; our ability to take the raw resources of our lives and to transform them into products of all kinds, serving every conceivable need. But we've grown lazy and self-indulgent. We've developed practices and attitudes in America that are not characteristic of those which made our country strong to begin with. Investors have grown reluctant to fund new products unless they happen to be in a glamor technology which can be almost instantly parlayed into profits, or on a grand scale multiplied almost cravenly. There is no interest in the slow, painstaking and patient kinds of investment which address enduring needs. Wall Street has grown predatory and Madison Avenue is bypassing all tangible values and panders mostly to illusion.

But I reserve my harshest criticism for we inventors ourselves! More and more of us are "welfare inventors," looking for a "quick fix" with a new gimmick. Many of us are simply not doing our jobs as inventors. Over the past two years, I've had the opportunity to meet and talk with hundreds of inventors and I detect more and more of a reluctance to get involved in anything more than "dreaming up ideas."

Perhaps we're all the product of our times. There's an alarming notion afloat suggesting that luck surpasses hard work. Lotteries are evident everywhere. There's a pill for everything. There is "professional" counseling for whatever is wrong. And there are other alarming revelations: American children are not learning; fully 13 percent of the high school graduates can't read their own diplomas. Japanese youth are mastering in high school what those of our children who do go on to college can't seem to grasp. I find more and more people ignorant about geography, history, science and a host of other subjects. We are becoming less and less resourceful.

Into this vacuum, various governments have thrown a multitude of resources to change this apathy and strengthen our capabilities. Throughout the United States, many new Small Business Development Centers have sprung up to assist inventors and entrepreneurs of all kinds. But there's a fatal flaw! Given the attitude so many inventors now have, these centers and other institutions are viewed as the complete answer to their needs. They'll take care of making me successful with my invention. They'll find me the money. They'll show me how to set up manufacturing. Or they'll find someone to buy my idea. Well it isn't so, and it shouldn't be. What they should be doing only is teaching inventors how to write business plans, showing them how to get a prototype made, advising them on marketplaces and how to find them and instructing them in entrepreneurialism.

The reason I say these things is because I am a firm believer in the strength of the free enterprise system and the wisdom of the marketplace. It is democracy in its finest form. And invention and innovation are the cornerstones of that system. Because invention is the most fundamental building block for new enterprise, it is a resource we must strengthen and support. Young people must be encouraged to question and solve problems. We must strengthen our own resolve. We must abandon the notion that someone else will do all the hard work and someone else will cure all our ills and someone else will provide all the answers. If we're to truly be successful inventors, we must be prepared to take initiatives beyond simply dreaming up new ideas. If we are not willing to put our own money and muscle behind our own ideas, if we haven't the courage to get our own small business started, why should we expect the state or federal government or anyone else to do it for us?

Such notions have failed dismally in places like Poland and East Germany. And Communist states everywhere—including the Soviet Union—are turning to the very free enterprise system which invention helped build. Is there any reason why we as inventors should abandon the very system we helped create?

It became apparent to me, when I tackled this second edition of the book, that a harder line or stronger suggestions were called for. Perhaps the first edition implied that inventing was too simple. Perhaps I didn't do my job well enough if I encouraged anyone to believe inventing was a "bed of roses." If I've become too strident in this edition, it is because I have seen so many good ideas abandoned simply because the inventor found that a little work was involved, and wasn't prepared for that. On the other hand, I've also tried, in this edition, to generalize less and get down to some real "nuts and bolts" suggestions on such things as financial data, determining product (and your own) potential, and making informed choices about what to do and where to go with your idea.

I have one final thought before launching into the book: As I point out in Chapter 7, there are flaws in our laws concerning intellectual property rights. These weaknesses affect us all. When an offshore company steals the intellectual property of an American company—for example a motion picture film, a computer or a medical technology—it costs all of us, and can perhaps even destroy an American industry. When someone pirates an independent inventor's discovery, and he or she has remedies so costly to enforce as to preclude any recovery, something's wrong!

Inventors, maverick as we tend to be, are not easily given to organization. We must! We should seek out and join local inventors clubs, find our common causes and make our needs known. Unless there is a constituency, lawmakers will not address system weaknesses. So get involved, and speak out! But first, get busy with that invention of yours and—do something!

Robert Park
Four Oaks, NC 27524

Introduction

The world needs good new ideas. And never before have we had such capability and the incentive for generating them. The incentives, beyond simple profit, must certainly include the priorities of better health and environment: a lessened world hunger and poverty.

In a truly Utopian world, people would certainly be freed from hunger, disease and poverty. Since life is instinctively dedicated to its own survival, any condition which threatens it provides insecurity and unrest. Hungry people revolt more readily than well-fed ones. And tyranny does not so easily survive in a contented culture.

If "necessity is the mother of invention," then invention most certainly is the patron of freedom. It provides freedom from hunger because of agricultural technology, freedom from pain and disablement because of the wonder drugs and ingenious new mechanical devices and instruments, and freedom from poverty because invention is truly the driving force that propels economic prosperity.

A tree in a forest or a mineral in the ground has only a diminutive value. But with ingenuity and inventiveness, we can turn the tree into a home or chair and the minerals into everything from crops, skyscrapers and fuel to beautiful gems. Money does not do that. Invention does. Make no mistake about it. And money goes with the action!

I am not suggesting that technology and invention can cure all our ills. Far from it. Atomic fission, despite its peaceful applications, has provided some frightening consequences and concerns. But so has the automobile, with our ever mounting highway fatalities. Likewise, many other inventions have lashed back, from miracle chemicals that poison to simple toys that injure a child. We cannot blame the invention; rather we must understand its appropriate use.

Despite the shrill outcries against the dangers in an improperly used infant formula, the political ramifications of "Star Wars," or even the oft-debated health hazard possibility of fluorescent lighting, power lines or radio emissions, new inventions will continue to spring from the imaginations of creative people everywhere. Political emphasis must be placed on the proper deployment of these inventions, rather than on banning the inventions themselves.

It is my firm belief that everyone with a reasonable curiosity about the world we live in, and an average imagination, can invent! Over the years, I have seen simply brilliant little inventions put forth by seemingly unlikely people. The hitch is that, often, the person with the idea lacks knowledge about how to translate it into a practicality. Sometimes it is a lack of technical expertise, often a lack of money with which to fund the invention's development and, frequently, the problem lies with failure to understand marketing and sales. In many cases, the inventor understands none of these requirements. She or he simply has a gem of an idea.

As a result, many do not pursue their ideas. While we all would like to become rich from a good idea or invention, I like to think that beyond mere profit, we have an obligation to our new ideas—to carefully consider what usefulness they have to others, and how we might make that a reality. If I conclude that an idea falls within a technical domain in which I am personally unskilled or uncomfortable, and I am not willing to become skilled and comfortable in the domain, I try to sell or even give the idea to someone more qualified than I. As many inventors do, I view my role to be as much custodial as proprietary.

In addition to the inquisitiveness which triggers discovery, an inventor must cultivate the courage to explore and test the idea. Don't be afraid to make a fool of yourself. I once did before a full audience of neighborhood kids. So sure was I that I could puff popcorn the way I thought wheat and rice are popped, I loaded a 12-gauge shotgun shell with the grain and blew a nasty hole through my bedroom wall! The kids cackled off down the street, declaring my insanity. But in fact, I had discovered something. I'd discovered how not to do it, and therefore was not obliged to repeat my mistake. That lesson taught me firmly that even a mistake teaches. Later, there were many trial and error or exploratory episodes in my inventing activity where I remembered that lesson and isolated a discovery by surrounding it with exploratory mistakes.

It has often been said that we tend to pick the wrong heroes. For example, we pick valiant warriors, sporting figures or movie stars. While their contributions should not be dismissed, we might also pay tribute to that multitude of inventors who have truly fashioned the world we live in today. Look about your own home, work or recreational place and consider what these men and women have added to your personal comfort and pleasure. Then if heroes there must be, surely you will include inventors among them.

Contrary to the notion that technology may be swamping and overcoming us, quite the opposite is true. More than ever before, through invention, we are truly awakening to the marvelous working of our universe. Each new discovery and invention leads us to incredible new vistas of understanding. One idea leads to another, and even the simplest discovery or invention has the potential for changing the world.

For example, curiosity about the germanium crystal's ability to detect a radio signal led to the invention of the transistor. Soon this was followed by the tunnel-diode. Ultimately these developments and others led to the modern computer. There are many now who say the computer has and will more drastically change the world than any other invention in history.

But, of course, not all inventions have such importance. So what! If an invention is but a trifle, yet amuses some child, or simply solves a transitory problem, it can still be worthy—and profitable. The market will judge. Who knows? A seemingly worthless gooey mess in the bottom of your test tube may become another Nylon or Scotch Tape or even the "Green Slime" so popular with kids today.

Inventing is also a lot like operating one of those little "Erie Gold Diggers" I used to encounter at the county fair. The objective was to turn the little handles and pluck some marvelous treasure from the pile of goodies. Sometimes I'd get lucky and snag the thing I wanted; more often, I'd take what I got, or nothing at all. And sometimes, even, I got something better than I had expected. But, in fairness to myself, I had to give it a try.

This book was written to encourage people like you to give it a try: first by asking the right questions, then finding the answers, improving the world for the rest of us, and—making yourself some money in the process.

Robert Park
December 28, 1989
Four Oaks, NC 27524

1.

Challenges and Pitfalls

We are all innately curious and manipulative creatures. Our propensity for toolmaking, carving, decorating and otherwise altering our natural surroundings distinguishes us from all other creatures. These insatiable appetites for change, alternatives and improvements have driven us from caves to the moon and beyond. No other factor has contributed so significantly to the shape and substance of our modern world as has this inquisitiveness and creativity.

The process is both self-perpetuating and ever-accelerating. Each new discovery suggests others. Each fresh plateau of advancement provides a vista of new possibilities. The superiority of nations is dependent on technology. The prosperity and well-being of a people is rooted in its inventions. Indeed, in our modern world, there would be few choices, little production, no profit, meager comfort, and a stagnant existence without invention. In fact, our world became modern because of invention.

Recognizing all of this, advancing cultures seek strategies for nurturing inventiveness and transforming creativity into useful products and processes. Realities which can advance our comfort, improve our security, diminish our suffering are also those which serve deep human needs and thus provide business opportunity, employment and new wealth. Invention is the blueprint for value-added. Unfortunately it is also one of our most neglected natural resources.

FOR INVENTORS, THERE IS NO SIMPLE SYSTEM

As important as invention obviously is to society, one would think there would be an orderly system for harvesting such creativity. There's not! Despite a burgeoning invention/entrepreneurial assistance industry in America, as we shall explore later, there is not a rational system for harvesting and commercializing new ideas and inventions. For one thing, creativity blossoms everywhere, not just in companies nourishing it. Most invention springs from everyday people like you and me. These ideas should not be confused with what comes out the doors of formal Research & Development activities; they are the ideas standing in the wilderness seeking an entrance into the system. These ideas originate with ordinary folks like you and me.

In addressing the problem of transforming inventions into profitable ventures, we enter a perplexing domain, dominated by misunderstanding and frustrating unknowns. At the very outset, it should be understood that by consequence of whatever motivation, an inventor has presumably already done some transforming of his own. Thus, we seek to profit from that which he has already transformed. And while

while this example might seem overwrought, it illustrates just one of many stumbling-blocks which so often separate the creative spirit from practicality. It is not a new problem; the Medici Popes suffered identical frustration with the temperamental Michelangelo.

Creativity flourishes in stimulating environments. The more eclectic our experiences, the more apt we are to create. And inquisitiveness and problem-solving talents can be developed through education and encouraged through recognition. Creativity is, however, not the major concern; there is an abundance of creativity within our culture and it is but the threshold. How do we get beyond that?

Ideas are neither good nor bad. We cannot call them up to order; often we cannot call them up at all! The same "flash of inspiration" that expresses triviality may alternatively yield a profound "discovery." But how will we know? The one certain test site for the value of an idea is that of the marketplace. Yet for most inventors, getting to the marketplace seems insurmountable. And think of how good ideas are simply wasted by neglect or circumstance. For example, when Hitler began his genocidal pogroms, Albert Einstein and his colleagues, all topnotch inventors, fled Europe to the United States to develop the technology which ended World War II. And when American investment neglected our threshold semiconductor industry, Japan seized the opportunity and today dominates the consumer electronics market worldwide.

To illustrate in another way, there is the lesson of the Hula Hoop. The Indianapolis inventor who could scarcely interest any "prudent" investment in her "novelty toy" ultimately sponsored an industry which purchased half the polyolefin production in the United States to satisfy the toy's market demand!

Conversely, fortunes have been squandered on "technologies" which address nobody's interest or are so poorly invented as to be non-competitive. What process then can seize the truly viable opportunities and banish the risks? What process can profitably transform invention into venture? First consider the "players" and their interests.

CASTING, EVERYONE

The Inventor. He or she is entitled to a profit, but what? For some inventors, recognition is the profit that appeals to their vanity. These individuals rarely comprehend profit as money; often they are motivated by altruism. Others are characterized as "welfare" inventors seeking desperately to loft themselves to riches. Many inventors bring little to the table save their idea and, moreover, are unwilling or unable to actively participate in the essential commercialization necessities beyond. Rarely does an inventor possess financial, production and marketing strengths in any combination equal to the specific need of the invention. Yet nature is capricious. Their ideas may be gems!

The Investor. The investor seeks profit. It can be derived from many sources besides new technology and therefore inventions must compete with alternatives for funding. Most investors seek short-term, high-yield response from their investments. But most invention requires long-term, high-risk and "patient" investment. Small Business Investment Companies (SBICs), designed to serve just such "seed funding" needs participate reluctantly with inventors. Banks almost never take part in the equity funding of new, invention-related ventures. "Angel" investors, who provide the majority of funding to individual inventors seeking commercialization, are almost invisible. Foundations are seldom directly accessible. "Revolving" public funding is frequently politicized and often encumbered with cachets.

The Community Developer. This includes the Chamber of Commerce, specific Economic Development offices, and quite frequently, an ad hoc consortium of local business people and political leaders seeking economic growth and prosperity for the community. Inventions which might provide the basis for a new manufacturing opportunity are candidates for consideration. But the economic developer has other choices too, such as that of encouraging tourism or enticing a profitably established industry from elsewhere into the community. Most economic developers have no protocols designed to evaluate an invention's potential for profit to the community. And an alarming few are prepared to offer or even identify a package of community resources necessary for proper commercialization.

The Public. Our interests are, of course, obvious. But most important is that businesses formed out of inventions provide employment and prosperity, and contribute immeasurably to our security and freedom. Still, these benefits often go unrecognized. We take technology for granted and we but reluctantly acknowledge the overwhelming contributions of inventors. They are more often perceived as "crackpots," "maverick," or worse. We select sports, music and cinema personalities as our heroes, seldom considering that without invention, their performances would be pallid or nonexistent. In our roles as consumers, we are fickle and unpredictable. We are a mysteriously elusive target for new products. Yet we respond to "New" unlike any other product feature.

The Glee Club! Here are all the fringe players that move in and out of every new product commercialization scenario. The patent attorneys, design engineers, lawyers, marketing people, packaging experts, media, machinists, suppliers, government agencies, insurance agents, employees, job shops, consultants, family members, and others. It's intimidating even to skilled business persons.

WHAT SHOULD I DO WITH MY INVENTION?

Here's where it starts. The inventor holds up his idea, and asks, "What should I do?" First, he asks himself, then his family. Cautiously, he mentions it to friends and perhaps co-workers. He thinks he knows what he has, but does he? Every fresh idea disguises sly truths. He looks for encouragement and validation of his idea. But who among this small group is qualified, other than by friendship, to judge? Still unknowing, innocent about the road ahead, naive about options, the inventor forms his own judgments and says, "Of course it's a great idea; I'd buy it!" He has at this point, of course, but fashioned a dream, an illusion. What was the question again? Oh yes, what should I do?

Mortgage the farm and go into business? Hire one of those "Invention Marketing" outfits that advertise in the back end of Popular Science? Go see my banker? Find a manufacturer who will buy me out? Go see a patent attorney? Look in the Yellow Pages? Join an inventors club? Hold a press conference? Make a model? What?

IT'S WHAT WE DON'T KNOW THAT TRIPS US

The painful fact is that most inventors simply don't know what to do or where to go. Many are encouraged first to apply for a patent, and after paying the legal and filing fees for this, have no money left for anything else. Still others pay exorbitant fees to the "Invention Marketing" organizations in exchange for nothing more than empty promises. Inventors write letters to manufacturers who invariably rebuff them—or don't even answer the mail. They seek investors, yet have no verifiable facts to support a proposal.

What the inventor seldom does do is this: Almost never does he go out to the marketplace to determine if there is an interest in his invention. Few seek a skilled evaluation of the product in terms of its funding, production and marketing requirements. Most inventors have never heard of "focus groups," never do a literature search of either the "art" into which their invention falls or the contemporary publications serving the marketplaces, and have little or no comprehension of packaging, public relations, commission salesmen, confidentiality agreements, pricing and markups, production processes, government regulations, trade shows, cost accounting or any of the other requirements of a successful commercial strategy.

ENTER THE HELPING ORGANIZATIONS

What are they? Depending on where you live, they're any number of things. First, there is SCORE (Service Corps of Retired Executives), the volunteer retired business persons organizations sponsored by the Small Business Administration. The inventor can set up a meeting with a SCORE advisor. Sometimes that helps. Often not.

Then there are the Small Business Development Centers, a "helping industry" now spreading throughout many states, and they're usually attached to the educational system. "They help inventors," says your friend at work. "How?" says the inventor. "Oh, they'll help you find money, and set you up in business or find a buyer for your invention." Well?

Well, not exactly! More often than not, they provide only palliative assistance. Most are staffed with people inadequately trained in the enormous challenges of new product introduction. Most have no truly comprehensive system for evaluation of either the product or the inventor's entrepreneurial qualifications. Few have any technical skills necessary for examination of new products or processes. Most truly have no working relationship or even comprehension of the network of resources available, and few have even identified prospective buyers or licensees of a new product.

Sometimes the inventor leaves with some "How to" literature provided by the Small Business Administration. He's often pointed to other resources where, for example, he can learn how to write a business plan. He's sometimes told about things like confidentiality agreements and patent office disclosure documentation. But almost never does he exit the process with a clear-cut strategy or even a definition of the options. Consider the inventor's dilemma!

WHAT DO ESTABLISHED COMPANIES DO?

At this point, it has become painfully obvious to the inventor that matters aren't so simple. This business of "Invent a better mousetrap" simply isn't true at all. Of course not! Consider what a well-funded, experienced organization such as 3M does, with exactly the kinds of ideas independent inventors have all the time:

At 3M and many other highly creative companies, a separate department has been set up to deal with new ideas. Those ideas that seem to have commercial merit are brought into this special domain, where free from the rigidity of the rest of the company's structure, they are stringently examined. If things look good, the inventor, a production person, and a marketing person are brought in from elsewhere within the company, a team is formed, and the entire resources of the company are made available to this team. Need a prototype? No problem. Need the names of some suppliers? You bet. Virtually whatever it takes. And a product is brought to market. It's nurtured, promotion techniques are tested, marketplace feedback acquired, changes made if necessary, production costs tied down, and ultimately, if it does well, it becomes part of 3M's catalog of over 50,000 products. And it works.

It works for them because two factors are present which are seldom present for the independent inventor. First, established companies begin the whole inventive process by listening to the marketplace. 3M's tape business began when a salesman discovered an auto paint shop in St. Paul was using surgical tape for masking—and leaving horrible tape residues on the finish of the car. He took the problem back to 3M and they invented masking tape! Second, an established company like 3M already has a thorough identification of all the varied resources needed to nurture and bring a new product to market. And yes, they have the money. And the sales force. And distributors. And brand loyalty. And accounting. Engineering, and whatever!

WHAT ABOUT OUR INDEPENDENT INVENTOR?

What's left for him? For most, after rooting around in the confusing turmoil of the process, it is discouragement. He becomes more confused than ever. As one inventor said, "The situation is characterized by obscurity!" And a painful realization dawns that if he is to do anything at all, it is going to require a major effort. It is generally at this point that inventors fall victim to an unscrupulous "Invention Marketing" organization and part with as much as $7,500 before realizing this is a blind alley too.

Let's remember, however, the inventor has a life embracing more than just his invention. A home, family, job, and other interests too. The invention is beginning to overwhelm all of these. It becomes his obsession. What he had once perceived as a shoehorn into riches becomes an albatross about his neck. The invention begins to loom all out of proportion. Objectivity goes down the drain. Not knowing what to do, he does nothing. Some, quite frankly go goofy! And what of the invention? We may never know!

BACK TO BASICS

To solve problems, we know they must be broken down into their simplest components, each of which is solved one at a time. However, all too often, in the mere process of compartmentalizing elements of a problem we lose sight of the primary mission. We become lost in the minutia. We neglect priority. We become easily stuck in serial activity, rather than parallel problem-solving. Much of this is the result of not knowing the lay of the land. If we had road maps, we wouldn't make so many side excursions into blind alleys. What would the map tell us?

It would probably be entitled "The Free Enterprise System." The journey would begin with this: "Find a Market and Serve It." Major features of the map would be "Quality, Service, Price." When we got to the junction called "Sales," we would find specific routes which involved things like: "Introduce Product," "Explain Features," "Transform Features to Customer Benefits," "Overcome Objections," "Ask For An Order." And the entire process of new product commercialization can be similarly characterized. There are definitive protocols for manufacturing, for packaging, for advertising, for everything necessary; there is little mystery about what's needed. Successful businesses practice them regularly, with quite predictable results.

So the inventor must have a plan, a map if you will. And like an AAA "Trip-Tik," it must be specific to his journey. And who provides such maps? Incredibly, nobody!

Why are there no "Invention Trip-Tiks"? Because to provide product-specific, entrepreneur-specific maps, an incredibly detailed underlying resource is necessary. The American Automobile Association, after all, could not provide "Trip-Tiks" if someone hadn't first mapped the territory. But the invention commercialization terrain has been mapped, so why not "Invention Trip-Tiks"? In large part it is because inventors don't realize they need them, and don't think they should have to pay for them! It is the most disastrous delusion of all. As Pogo says, "We have met the enemy and it is us!"

Back at the Small Business Development Centers. Here, if properly trained counselors had the resources, they could provide the "Trip-Tiks." But mostly, they are not properly trained and they don't have the resources. (You can't properly train someone in the use of a resource that doesn't exist!) What they do present, if anything, is a general map. The inventor is left to pick his own route.

All of this is, of course, a poor substitute for what companies like 3M or Hewlett-Packard do. And most discouraging of all, it is just such competition that our inventor will be facing in the marketplace.

THE DISASTERS Consider the kinds of problems inventors have which might have been eliminated if serious resources and skilled counseling were available:

> An inventor has a new product fully engineered, market-tested, attractively packaged and ready for production. But he lacks funds! There are investors looking for just this kind of opportunity but the inventor doesn't know how to write a business plan with which to approach investors he can't even find!

An inventor develops an ore recovery process that could save millions. But it's environmentally hazardous!

An inventor spends his life savings on a technology only to discover that an alternate technology offers a cheaper, better solution!

An investor funds a new venture only to discover the product will require liability insurance equal to his total investment!

An inventor spends precious funding on packaging for a great new consumer product. But because he neglected to include bar coding, no major retailer will buy!

Such are the treacheries of the inventing business. And we have no system for dealing with that? It is tantamount to operating a bank without computers!

THE OBSTACLES   There are a number of pitfalls; not the least is the inventor. Many really don't want to succeed. Some are like authors who never publish; they are fearful of the dazzling brilliance of the marketplace. Many are so attached to their delusion, they'll listen to nothing else. For this too, there are alternatives.

There is a certain "status quo" benefit existing businesses enjoy by leaving the present non-system intact. This neglects the possibility, however, that an inventor's procedural ignorance may deprive all of us of the benefits of a truly meritorious invention. It may also deprive us of new businesses and jobs.

Many communities take a peculiarly guarded view of new business development. One must probe beneath the surface to understand that a complexity of sentiments and special interests is at play. Many communities have been hurt by past schemes. And it must be remembered that politicians and other civic leaders live closely with neighbors. Disasters are never welcome and difficult to defend. On the other hand, if these leaders were able to access carefully evaluated, properly managed, mystery-divested new product possibilities, which could provide the foundation for a new industry or offer opportunity for existing businesses, it would be another matter entirely.

The existing infrastructure has its own self-interest and in bureaucracies, there is all too often a self-preservation instinct having little to do with the mission of the organization! And it is politically expedient to organize around the press release or TV "Bite," rather than tangible actions and the many individually small accomplishments which comprise the elements of success.

INVENTION IS VITAL TO US ALL

One of the most important elements of our nation's great success is our free enterprise system. It placed democracy squarely in the marketplace. We looked to the marketplace and invented products it needed. Of course we had the optimism and courage that all our other freedoms provide, and we had a continent rich in resources and the confidence to take risks. And the system has worked very well indeed. But there are disquieting portents.

America is losing ground. It is scarcely necessary to recite the frightening statistics. It is alarming enough to witness the dissolution of our small towns, the plant closings, the hopelessness of "dis-employed" workers. Where did we go wrong? We, the nation that innovated our way into the strongest, most affluent nation in the world.

Clearly, it is a new ball game. As one economist put it, "World War III has already begun, and it is an economic war." We must return to the most prodigious resource we possess, that of our creativity and inventiveness. And why not? We have even better tools to work with now than did our forefathers. We must not foolishly experiment with "socially apt" programs which shield competitiveness from the reality of the marketplace. We should provide incentives, and above all, we should stop talking about it and start doing it while we still own our own country!

Just as important, let's discover how we can transform your brilliant gem of an idea into the practicality of profit.

2.

Inventing Smart

If an inventor is to make money, his or her idea must be exceptional. A clever idea is just not enough. The world doesn't need more faulty mousetraps.

Unfortunately, cleverness and practicality are not necessarily synonymous. While frivolous inventions sometimes make money, more often they do not. Fortunately, there are measures we may apply to a clever idea which enable us to make some predictions. Consider these examples:

Recently, an inventor unveiled a novel theft-proof salt and pepper dispenser—a hinged, frame-like device which locked the cellars into a position enabling dispensing of either or both salt and pepper. The whole apparatus was then linked to the table with a chain. On the surface, the inventor would appear to have addressed a need, to keep salt and pepper dispensers from being stolen from a cafe table. But had he? Something's wrong! Let's examine it.

First, how many salt and pepper shakers are stolen from cafe tables? Is chaining them to the table appropriate? Or is identifying and ejecting the patrons who steal them a better solution? Does chaining them down affront patrons who don't steal? And would prepackaged salt and pepper packets solve the problem with equal or less expense? If the thieves can't steal salt and pepper, will they steal catsup? Or silverware? There lies a thought.

If you examine what people steal at a restaurant table, you'll find it's seldom salt and pepper. The incidence of theft is no greater than the mischief of pouring salt into a sugar dispenser, or vice versa. What, in this day and age, people do steal from a restaurant table is spoons! The inventor has addressed the wrong need.

ANALYZE THE PROBLEM

We have several premises here. First, an inventor must address the proper need. Second, the inventor must address it in the best way. Third, he must not affront traditional sentiments.

Let's look at the latter first. Many inventions, while useful in every other regard, have a tendency to require a user involvement which is unseemly, possibly even embarrassing. A classic example might be the fedora hat which unfolds to become an umbrella. On the surface, it looks clever, but consider the kind of involvement required of the user. Since rain is usually accompanied by wind, and the unfolded umbrella-hat has a tendency to sail, an additional provision must be made for affixing the apparatus with a strap beneath the chin.

TWELVE OF MANY POSSIBLE CONSIDERATIONS FOR INVENTING SMART

1. Listen to the marketplace.
2. People don't necessarily want what they need.
3. Understand and address what people want.
4. Do it better, at a lower cost.
5. Solve people's problems.
6. Simplify, don't complicate.
7. Make features obvious.
8. Make features suggest benefits.
9. Seek to develop independent rather than dependent inventions.
10. Consider up front both manufacturing and marketing necessities.
11. Cleverness and practicality are not necessarily synonymous.
12. Don't require the user to specially accommodate the invention.

Second, the wearer must also resort to some gymnastics, balancing the device on his head in the same manner he would manipulate a conventional umbrella toward gusts of rain. Parenthetically, there is a certain elegant flourish about the conventional manipulation of an umbrella. Most important—and this is the main point—the user does make a spectacle of himself! He becomes not distantly removed from a kid wearing a propeller-driven beanie cap. And people do not ordinarily like to make spectacles of themselves, least of all in pursuit of legitimate objectives.

Working back to the second proposition, using the same umbrella-hat illustration, does this solution address the need best? Does an umbrella, even? Does an umbrella protect one's feet and trousers? Therefore, is a raincoat suggested? Or a better version of a raincoat? This leads us to the understanding that simply because a device is clever, it does not necessarily mean it is properly addressing the need. As stated earlier, cleverness and practicality are not necessarily synonymous.

And now for the first proposition. Since the umbrella-hat is associated only with a man's fedora, it neglects whatever need half the adult population—the women—might have, and since kids aren't wearing fedoras, it neglects still another market segment. The umbrella-hat is simply not very fully addressing the need.

The creative process prefers an unfettered climate; the flash of genius is often nothing more than the recognition of a need. Thereafter, the process too often deteriorates. We are trapped in the allure of the discovery and abandon the tedium of finding a best solution. Thus, many inventors simply compound elements to solve a problem which results in little more than an embellishment upon existing methodology.

If, having determined in a general way that there might be a need, the inventor should posit as many solutions as possible. Here, the imagination should run free. Every possibility, no manner how abstract or ridiculous, should be examined. This is truly the core of the inventive process. It takes little regard of rationales or even consequences; it focuses simply on how.

If there is a single, simple rule for this how step, it is that the answer to every question invariably lies within the question; the solution to every problem lurks somewhere within the problem itself. Thus, identify the true problem and ask the right questions. This is important because the how step involves a wide-ranging survey of potential relating factors and too often the inventor strays irretrievably from the primary focus—the problem.

STUDY HOW NATURE COPES

Study nature. Many of us, so isolated from the majesty and complexity of it, overlook the fact that nature has developed extraordinarily ingenious methods of coping with problems. In fact, scientists often discover the coping mechanism before discovering the reason for it. Just as every artist is counseled to refer first to nature, so should every inventor. We must look not at the forest, but the tree!

While this would seem to violate the dictum that an invention must not exist in nature, it isn't so. Consider the 3M product "Scotch Guard." While we don't know the deliberations which went into that inventive process, it might be suggested that in a certain way, this product addresses the same problem the umbrella-hat did, but from an entirely different angle. There are equivalents in nature, notably in waterfowl who employ a similar water repellent to stay afloat. Part of the cleverness of the "Scotch Guard" product, however, is that having addressed the problem of water, it also went on to address that of stains. And thus the product solved problems beyond that of simple rain.

At some point an answer will be found for a problem. But is it the best answer? An inventor must never accept the first whimsical answer that pops to mind. He should put that answer aside and look for other answers. Put those aside also, and look for still other answers. Strive for simplicity and functionality. Eliminate compounding methodology. Think in terms of coordinating functions and dovetailing elements. When all that is done, and perhaps at every stage of that test-and-try process, the inventor can make determinations about how adequately each solution addressed the initial problem. Sometimes it will be found to only partially address it; other times it will be found (as in the "Scotch Guard" example) to address other problems as well.

NEVER ACCEPT THE FIRST ANSWER

There's an extremely important reason for this test and retest procedure. Remember that your identification of a need is the first step, and what you have come up with is your answer. But what if someone else, having been pointed to the need by your invention, probes further and comes up with a still better answer? This, in fact, happens all the time, which leads to the unhappy scenario in which the inventor cries foul and contends his idea has been stolen from him.

In truth, the inventor did not work hard enough in the second stage of his inventing; he lazily accepted a solution which was inadequate. The painful fact is that often the solution the inventor did arrive at simply provides an excellent starting point for another inventor to make an improvement or discover an entirely better answer. Not only did the original inventor point to a need, but his invention pointed directly to a better invention.

Satisfied the invention does most cleverly solve a problem, most substantially address a need, the inventor must then look at the original third premise—does it affront the user? Too often, inventions require an inordinate accommodation by the user. The umbrella-hat is one example, but there are many others. The cure becomes worse than the ailment.

Let's "invent" an illustration: The flash of genius is that an electric lawn mower (or hedge clipper) could be improved with an attached light. Remember, we already have the electric power there. This would enable the user to finish work in the dusk or darkened hours and thus extend the usefulness of the main product.

THERE ARE A LOT OF DUMB INVENTIONS

So the inventor devises a light, and appropriately incorporates it into the lawn mower. What's wrong with that, you say? What's probably wrong is that the user then attempts to mow his lawn late at night, to the utter distress of neighbors who have better things to listen to. The invention, if it succeeds at all, succeeds also in making the user a nuisance. Unfortunately, there are all too many of us who buy gadgets of that kind without considering the real consequences.

Consider, for example, the proliferation of kitchen utensils. It seems that one of the most popular strategies many companies have for new product development is in ever-more-narrowly addressing needs in the marketplace. Hence, there are virtually hundreds of products— they come and go—which essentially address needs already served by common

---

INVENT WHILE YOU SLEEP!

Researchers now believe that during REM sleep (so called because it is accompanied by rapid eye movement), the brain processes information which has been temporarily stored during the preceding wake period. During this process, "data" are sorted, some added to permanent memory "sites," and other information discarded as irrelevant. The process is analogous to that of a computer updating a current disk file.

Other studies suggest that the creative process is very much a sorting process in which the brain shuffles new and old information together to yield a sort of "hybrid" new thought—something caused by the "rubbing" of ideas together.

Many creative people have told how they went to bed with an idea or problem in mind, only to wake in the morning with an answer. Possibly the creative process works best in the subconscious, and we should turn more of our inquiries over to our sleeping self. In any event, pleasant dreams!

utensils such as knives, pots, and traditional ware.

The reason most of these products do not survive long is that ultimately the consumer realizes the gadgets: (1) Take up a lot of already crowded drawer or counter space, and (2) they require a lot of extra cleanup and dishwashing. Common sense returns and the gadgets wind up in the next garage sale.

If, on the other hand, an inventor addresses the same needs with a view to addressing function and simplifying, there is an opportunity for a product with considerable longevity. In short, a product must survive its own novelty to endure.

A good example in this area is the toaster-oven. One device; a multiplicity of uses.

FRIVOLITY IS NOT NECESSARILY BAD

All of this is not to suggest that frivolity and embellishment are unacceptable. Many products succeed, at least for a time, satisfying a whimsical need. Fashion and style changes dictate a continuous marketplace for inventive applications.

How does one focus on a genuine need and pursue an inventive solution? Here's a wild card: Recently, in idle conversation, someone posed the question: "I wonder if a mosquito could transmit AIDS?" It was a horrifying thought! Coincidentally, on a talk-show that evening, the same question was posed to an expert in that field. The answer he gave was that no, the virus was apparently disabled by the mosquito. Disabled?

"Good grief," said the first conversationalist, "Somebody had better be looking at mosquitos and finding out how they disable the AIDS virus. Because maybe, just maybe, if they can figure out what the mosquito does . . ."

Outlandish? Consider this: The drug held out as the only currently workable treatment for AIDS is AZT. It was originally discovered in and derived from herring sperm!

LISTEN AND OBSERVE

And this is how invention works. One listens, observes, and perceives the needs, whether they satisfy only a whimsical desire, or a catastrophic one like a treatment for AIDS. Then without regard for other people's strictures or one's own innocence, you tackle it.

But lest you stray too far from your mission, there are some guidelines for the inventor, and if they're carefully observed, your chances for success are immeasurably increased. Whether you already have that great idea, or are trying to figure out a new one, consider these:

1. Listen to the marketplace. Invent what people want. Look for the needs and come up with solutions to problems.

2. Invent it smart. Invent and reinvent. If you don't, your idea will only teach someone else how to do you in. Simplify, don't complicate.

3. Create features that serve up benefits. Features are what you offer; benefits are what people buy. Make benefits as quickly and clearly obvious as possible.

4. Find a market and serve it with quality, price, and service. These factors are all ultimately far more important than the transient novelty of your idea.

5. Act as if. Put your idea into a new business where as owner, you can put yourself to work turning the idea into a reality. Set goals, and establish priorities.

6. Don't think about it, or talk about it. Do it!

3.

The Real World

The popular misconception is that if you have a great idea, success is just around the corner. Inventors go to their graves clinging to that delusion, and blindly fail to recognize that there is a great deal more to an invention than just an idea.

One way to view the situation is to begin by thinking of the idea as a recipe. Have you ever been to a bake sale and found anyone selling recipes? Of course not! People don't buy recipes at bake sales, they buy pies, because they can see, smell and even taste pies, but not recipes.

So it is with the process of inventing. The idea is the recipe, but the invention is the pie. If you're going to have any luck at either a bake sale or with your invention, you've got to get cooking!

The idea, then, must be turned into a reality. More than that, it must be transformed into something that provides benefits and value to others. If you were to make and sell the product yourself, that means turning the idea into a tangible something that people will be attracted to and will buy. If you contemplate selling your invention to a manufacturer or licensing it for royalty, you must be able to present not just an idea, but a "profit center."

Many inventors treat their brainchild in a sort of ad hoc manner. They "kind-of-sorta-maybe" it to death. One professional characterized typical inventors as people who "ready, aim—aim—aim—!" What's missing?

DON'T JUST THINK ABOUT IT, DO IT

What's missing is action. Why? I think because often the inventor doesn't know what he's committing to. He or she has a foundling invention, and isn't sure if it will turn out ugly or beautiful. The plain truth is many are reluctant to admit parentage of an idea which others may scoff at. Yet, you can't even discover an invention's worth until you make the commitment to discover it. And discovering it is not for the lazy.

The inventor may fashion a crude prototype and show that around. About this time, he begins to wonder what manufacturer will buy him out. And perhaps he writes a few letters. The response is typically underwhelming, if they bother to answer at all.

Sometimes, at this point, the inventor approaches a bank or investor, and usually with no luck. Sometimes, he's told he must get it patented before any money can be found. Off he runs to the patent attorney.

What happens next is generally sad, because already the inventor has the cart before the horse.

Despite my general respect for the technical competency of most patent attorneys I've met, I've yet to meet one who pronounced an idea unworthy of a pursuit of intellectual property protection. Thus, hearing this "good news" from the patent attorney, the inventor digs deep and begins the expensive process of getting a patent position.

What is sad, indeed tragic, about this is that the inventor hasn't even figured out what the potential commercial merit is of the invention he seeks so diligently and expensively to protect. It's about like locking up a mirage! Worse still, the inventor seldom is given to understand exactly what the protection is that he will be getting with a patent. Curious?

INVENTIONS, PATENTS, AND PROMOTERS

All a patent grants the inventor is "the right to exclude others from making, using or selling" the invention. Read that line carefully. It is saying simply that this invention is yours and if someone tries to horn in, you can sue. YOU, not the patent office. Do you have the money to launch a lawsuit? Against a large, mean-spirited predator company? Think about it.

It is perhaps at this point the frustrated but undaunted inventor falls victim to an unscrupulous "invention marketing" firm who offers to sell his invention to a manufacturer. I've heard of fees of up to $12,000 for such services, which usually consist of a generalized boiler-plate report, and some rather crude specification sheets mailed to a very few companies. Of the thousands of inventors I've communicated with, I've met none who ever profited from such an excursion. Caveat Emptor!

And so what we've seen is wheel-spinning. The inventor has done all manner of things trying to validate his idea, but in fact has simply been flattering his own ego! Indeed, the patent office itself has been called by some "the biggest vanity press in town." And remember that despite whatever the patent attorney may tell you about your invention, his competency is directed to the technical prerequisite of your idea for patentability, not its commercial merit.

THINK MARKETPLACE

What many inventors fail to do is examine the idea in terms of marketability and profitability. Will people buy it, and can money be made making and selling it?

Those are the things an inventor must determine, and spending a lot of time soliciting unqualified opinion about the invention, endlessly speculating about unknowns, seeking protection for the still untested concept, and talking it all to death simply don't advance the idea to profitability. Worse, these misdirected engagements take attention away from the important steps that must be taken.

Before you invest in a patent, or make any major commitment to your idea beyond a prototype (to establish if it even works!), you have to answer these questions concerning marketability and profitability. We'll examine those questions carefully, and work through them in detail, but first, let's get an attitude tune-up and a look at the system that is actually out there.

If an endeavor is going to work out for you, you must maintain a sense of perspective about your mission before you jump with novice enthusiasm into deep water. Let's take a close look at what makes an appropriate attitude, because it will be important to you on your journey ahead.

ATTITUDE IS EVERYTHING

Who can forget the nursery story of the little engine that could? Anticipating the steep grade in the track ahead, the train began to huff and puff, "I think I can. I think I can. I know I can. I will, I will, I will!" And of course, the little engine did.

David Riesmann, in his classic book *The Lonely Crowd: A study of the changing American character* (Yale University Press, 1950), suggests that children brought up on the "Little Engine" and "Jack and the Beanstalk" nursery stories or the Biblical "David and Goliath" legend tend to develop a greater sense of self-reliance and internal fortitude than children reared on modern fare such as "Superman." In reading the traditional tales, the child develops a keener sense of self-power. In the more modern story about Superman, however, ordinary Clark Kent's extraordinary power as Superman is invoked from a supreme external source, not from within himself. The child attaches this conclusion to his or her own personal capacity, then all too frequently damns the "Fates" for failure actually resulting from his own incomplete or nonexistent efforts.

If Mr. Riesmann's assumptions are correct (I believe they are), the kinds of stories we tell our children may be one cause for some of our society's flaws. Our blind submission to the corporate image and to "overwhelming" technology, and our lack of faith in our personal capacity for achievement, are but two examples of such defects. In other words, stories, whether for children or adults, may dictate either a tendency to be "inner-directed" or "outer-directed." Our attitudes toward accomplishment become strengthened or weakened.

Attitude is defined as a state of mind. We may have bad, good or indifferent attitudes. But the kind of attitude I believe most appropriate would best be described as heartily realistic. I refrain from the word "optimistic" because it suggests the possibility of delusion, and I disdain "pragmatic" for being too rigid.

I once worked with a man whose job was to buy, sell and trade millions of dollars in foreign commodities. The responsibility was enormous and it called for a cool head. I discovered he measured his own temperature and blood pressure once an hour! He said he was plotting his metabolic and emotional curve to determine whether he was in an optimistic or pessimistic frame of mind. "Oh, so you can buy or sell when you're in a good mood?" "Oh no," he replied. "When I'm in neither a good or bad mood, but an objective and realistic one." I don't know if it worked, but it sounded as if he was on to something.

Or, for another example, when you apply for a job, the most important qualification you can offer is a good attitude. No matter how well qualified you are in every other respect, if you have a negative attitude, no one wants you! That seems simple enough, but it is positively amazing how difficult it is to apply that simple truth to ourselves.

THINK POSITIVE

I don't mean to parade all the benefits of "positive thinking"; the public libraries abound with excellent texts on the subject. But I'd like to encourage you to consider these realities:

> There is no conspiracy "out there" designed to prevent you from being successful.

> The world's "dumb indifference" is not a rejection of you or your ideas but rather, with its necessary preoccupation with its own more pressing concerns.

> For the most part, you can do it if you simply get busy and do it. Not think about it, or think about doing it, but do it.

But how does one get and keep the right attitude? By self-trickery, if necessary. But, most importantly, by gaining the victory one step at a time. Define the problem. Break it up into its smallest components. Prioritize your list of components. Identify what needs to be done to solve each of those component problems, one at a time. And then do them! One by one.

DON'T UNDERESTIMATE YOUR OWN CAPABILITY

Often, you will identify a particular element in your list that needs to be done, and you not only have never before done such a thing, but haven't the faintest idea of how to do it. Usually the solution is as simple to acquire as going to the library and studying a book on the subject. Or you may want to ask friends or, possibly, suppliers for help. Then you try to do it. Perhaps, the first time, you will be unsuccessful. Try again. And again. "Oh, that's too much work," you say, or "I could never learn to do that!" Well if that's what you believe, you'll probably fail. But if like the little engine, you think you can—you know you can—you will, you will, you will, then you will!

Of course you can retain someone else to do your job for you. There's a place for that. But by and large, you will be surprised at how many tasks you actually can perform yourself. It's simply a question of getting the right attitude, putting your mind to it and doing it.

One of the many framed legends I keep around my office says this: "It isn't the mountains ahead that wear you out, it's the grain of sand in your shoe." If every obstacle is treated as though it were but a grain of sand in your shoe, it can be dealt with, a grain at a time. And when it comes to the mountain, even that can be conquered, a step at a time.

YOU HAVE TO START TO GET THERE

Often, the transition from identifying the step to actually making it is of nearly cosmic proportions. A peculiar inertia, born in part by self-doubt or fear of the consequence, tends to root us to inaction. It is an almost adolescent affliction for me, locking me in the warm embrace of a mere daydream and inhibiting my departure into actual accomplishment. Fear of failure or rejection is no doubt part of it. But it can be overcome with action!

Then there is this matter of laziness! Actually, there are two kinds of laziness. The first is that tired, listless do-nothing feeling we all experience when overtired or physically run down. In part, it can result from poor personal habits; you aren't eating right or getting proper exercise and sleep. To correct that kind of laziness, you must go to the source. Take care of yourself! Get off the roller coaster. Try to avoid the excessive ups and downs, the needless bursts of energy, and measure your pace. It's amazing how this kind of discipline can smooth out your life and your attitude.

The other kind of laziness is brought on simply by your state of mind, your poor attitude. Things don't start out right for you, you get some bad mail or none at all, or you break your shoelace first thing. Then your whole day goes "down the tube." Nonsense! These kinds of things happen to all of us all the time. It's how you react that makes the difference. You deal with such frustrations by not dwelling on them needlessly, by not whipping and punishing yourself with thinking about them. Think of something else. Concentrate on the positive needs. Your attitude and self-motivation will soar. It may take some practice, but you can do it—if you do it.

I mentioned self-trickery above. While that may not be the best term for it, there are certain techniques that closely characterize what I mean. When all other self-motivation fails, I trick myself into doing a job by arranging the task so that I have no choice but to do it. For example, if I have to repair a broken shelf in the closet, I clean off the shelves and move the articles out into the room. Then, as quickly as possible, I remove the broken shelf, and I'm stuck! To get things back in order, my best choice is to finish the repair.

I learned a variation on that stunt years ago as a detail salesman for a candy company. As usual, getting started was the toughest part. I finally developed the habit of making an appointment by phone with a store out at the farthest section of my territory for the earliest hour possible the following day. Once I'd done that, I had no choice but to keep the appointment. Instead of haphazardly working my way out into my territory, I started at a far point in it, and worked my way back home. It was a kind of self-trickery that was really my technique for self-discipline.

DON'T BE AFRAID YOU'LL MAKE MISTAKES

If you've ever taken a piece of clay in hand and begun a sculpture, you know that making the first few manipulations are the most difficult. But once you start the process, each effort leads to a succeeding one and, gradually, the mass yields to some identifiable form. Not every stroke or indentation will be perfect, nor does it have to be. Each and all are correctable. But the only way to overcome your fear of making a mistake is to risk action and to believe in your own capabilities.

Nearly everyone has heard the adage about the bumblebee being technically unable to fly, but not realizing this of course, it flies anyway! Or perhaps it is that like the little engine, it thinks it can and does. Think what a positive attitude can do!

CONSIDER VALUE POLITICS

But now look at the system you'll be up against! First there's the matter of what I call "value politics," something inventors often fail to grasp.

Mankind has developed a sophisticated method of accounting for value of all kinds and that is, in a word, cash! What has this to do with your invention? Altruistic considerations aside, it has precisely to do with the matter of exchanging your invention or idea for the more uniformly identifiable value of cash. And here is where you run into "value politics."

The first thing to remember is that the individual holding cash has an automatic superiority in bargaining strength. The only way to equal the bargaining odds is to convince him that combining his cash with your idea will yield a profit. After that, it is simply a matter of haggling over percentages.

EVERYONE ACTS IN HIS OWN SELF-INTEREST

The individual who has the gold or cash must be persuaded to part with some of it to obtain your invention. Why would he or she want to part with the divine security of cash? In the marketplace, at all levels between your invention and the ultimate consumer, the only reason any participant will trade cash for your invention is profit. Only in the realm of conspicuous consumption does the profit motive falter slightly. But even there, profit may be tabulated in more aesthetic terms such as "pride of ownership," the "trappings of power" or other ostentatious coinage. It is still profit, of a kind.

The bank or other financial resource which may fund your invention, the supplier who furnishes equipment or materials for production, the employees who work at its assembly, the wholesaler and retailer who purchase the product for resale, and the final consumer all rightfully expect to "profit" from their involvement in your invention. Profit then is the fundamental issue, and value politics has to do with the art of translating value from one form to another; in your case, how can your invention be translated from its original form into another value form (cash!), at a profit to those involved in the transaction?

This digression into the economics of inventing may seem improbable. The inventor, however, too often overlooks this basic concern and must constantly tug himself back to fundamentals. Novelty, pride of authorship, altruism, or mere delusion blind many to principles of value politics. Many inventors, stricken with the "earth shattering potential" of their idea, fail to realize that this "potential" is far too elusive a value for the moneylender or merchant. Their interests, if any, remain those of profit!

The inventor must recognize also that since cash is a known and far more universally definable value, its transposition into funding for your product development constitutes a much more perilous, or at best, undefined value. Persons who take this risk are entitled to profit.

From the standpoint of your bargaining position in these matters, then, it is important to remember this most important fact:

Invention is the critical catalyst between virtually all natural resource and profit derived in the marketplace. Without invention and technology there can be no cultural advancement, no manufacture and no profit. Without it, we'd all still be in caves!

INVENTIONS ARE BARGAINING CHIPS

This is what you as an inventor are all about. This is the cornerstone of your bargaining strength. For communication to take place between the cash holder and yourself, you must define the rate of exchange. As in international monetary marketplaces, a bargaining process is involved. The value relationship between your commodity—your invention—and the cash is determined by how well you can convince others about your product's profit-making utility. The further you are able to progress into the marketplace with your invention, the greater the value attached to your endeavor. A raw idea has only an estimated value, a prototype another, a production sample another and a highly popular consumer-accepted product, another still. It behooves the inventor to invest as much of her own time, energy and money into the germinal stages of the project if only to enhance her bargaining strength.

UNDERSTAND THE OBSTACLES

Then there is the "Corporate Moat," that seemingly impenetrable wall separating the inventor and his idea from a starring role in the boardroom. Many inventors have the delusion that their "Mousetrap" will be quickly snapped up by the highest bidder among a roster of competing corporate giants. If General Motors can't use the invention, Chrysler will. Right? Not necessarily! Since many inventions are dependent on a market already established by other technology, sale or licensing of the product is generally at the mercy of major market-dominant firms. And there they sit; they seem haughty, elusive, and unresponsive to your dream.

The inventor seeking funding will more than likely confront the same kind of aloofness among financial firms, and down the line, among chain buying committees and even supply organizations. So it is important to understand something about this so called corporate moat, and how to bridge it, or perhaps to consider building your own castle.

Most companies have active in-house product development programs. When approached by an outside inventor, the company is reluctant to respond for several reasons:

The company may be engaged in research and development of an identical or similar product, and does not want a costly entanglement with some stranger who will want a share of profit from that development. Both you and they may have invented the same thing!

The company may prefer to work with their own staff, because the staff has already proven its capabilities and works toward product development fitting specific corporate goals and marketing strategies.

The company may be particularly sensitive about an outside entanglement infringing on confidential areas, such as trade secrets, government or other classified relationships. Bluntly put, for all they know, you may be a spy! They are merely protecting their own trade secrets.

Inventors are often perceived by managers as "crackpots." And let's face it, a few are crackpots, even if their ideas are relevant and practical. The creative, "undisciplined" inventor may seem out of place in a structured corporate environment, and a manager may be reluctant to work with an "outsider."

However, sometimes the inventor has problems in the corporate world for reasons less valid than those listed. The structure of a large corporation often conceals apathy or incompetence at many levels of management. Such employees may be indifferent to new ideas or may prefer to "play it safe," to ensure their own job security. Or your inquiry may simply be lost in a stack of papers on the desk of some overworked or under-motivated employee.

INVENTION LICENSING IS A DIFFICULT PURSUIT

Our purpose here is not so much to denounce the establishment, as it is to understand simply that it does not favor the independent inventor. Licensing or selling your invention may be one method of reaching your goal; however, it is probably not going to be the easiest or safest course. At its earliest stages, your invention's ultimate value is poorly defined. It has not yet survived the wisdom of the marketplace. Your bargaining position is therefore weak. What other choices does the inventor have? One option is to make and sell your own product. Believe it or not, it's done all the time!

4.

Plotting a Course

If you're fortunate enough to have come up with an idea that looks good and seems exciting, your questions are probably "what shall I do?" and "where do I go with it?" Often an idea falls into a manufacturing category unknown to you, or into a marketplace you are unfamiliar with. While you may have invented something in an area you think you know and understand, usually inventors find they actually know very little at all about the complete picture.

For example, a painter develops a device for cleaning paint brushes. He understands the problem of cleaning brushes, and the mechanics necessary to fashion his device. But suddenly he discovers there is a whole lot he doesn't know about the paint or paint brush business. He has little idea of who's making brushes or paint, what the channels of distribution are between those manufacturers and stores that sell the products, and indeed, isn't really knowledgeable at all about how paint and brushes are made and sold.

Most of us are that way; we work in fields in which there are special niches, and while we may understand what we're doing, we're often ignorant of what the worker next to us does. I know a very skilled electric power lineman who has been climbing utility poles for 20 years and when I asked him what kinds of trees the poles came from, he hadn't the foggiest notion! In another instance, when I was developing colored popcorn and explained to a friend what I was trying to do, he responded with, "Is that seed a monocot or a dicot?" It floored me, because the truth was, I didn't even know what the terms meant. Intent as I was on the process, I had neglected to fully understand the nature of the thing I was trying to process.

At a recent innovation workshop, the principal speaker was the head of new product development for one of America's largest and most innovative companies. Throughout his talk, he used the phrase, "Go to the literature," and what he was trying to instill in the audience's mind was the extreme necessity for understanding the whole picture. In fact, inventors very frequently find refuge in just one small facet of their inventions, and neglect the world surrounding it.

If an invention is to be successfully commercialized, the inventor must discover where it fits into the larger scheme of things and how. How will it fit into manufacturing? Where is the marketplace for it? And everything in between, before and beyond. People working in research or academic surroundings invent things and have no comprehension of the marketing necessities for what they've invented. The same is true of people working in marketing or sales; they have no understanding of what it takes to make the thing they've invented. The answer is: "Go to the literature!"

DEVELOP A PERSPECTIVE

One of the principal attributes of a good entrepreneur—something you'll be forced to become to profit from your invention—is the ability to grasp the big picture. One cannot understand the relevance of the invention without understanding what it really is and what it is supposed to be relevant to. Understanding the big picture grants the inventor a perspective on commercial merit and allows for the development of a plan with goals and priorities. Without this complete knowledge, the chances of success are slim indeed.

Many inventors avoid this process, as if ignorance would shield them from failure. It is kind of a "what I don't know can't hurt me" attitude. What if someone else has invented the same thing? What if there's something better out there? What if my competition is a large company I can't compete with? What if people won't like my idea?

Or how about this: The inventor has developed a product that is manufactured by injection molding plastic. But he knows nothing about that particular manufacturing process. He says to himself and others, "Oh I don't bother with those kinds of details, because I can pay the supplier to handle it for me." Rubbish! If he doesn't understand the basic manufacturing requirements for the product he wants people to buy, he's engaged in an utter neglect of responsibility to the public and his own success with the invention.

Here's what it boils down to: *You're not going anywhere if you don't have a goal or destination; you can't pick a destination if you don't understand the choices and you can't make choices if you don't study the possibilities and limitations!*

OUTLINE YOUR STRATEGY

Let's begin then with an outline for developing a strategy. First, there are the major choices and the needs:

1. Should I seek intellectual property protection? Is it patentable? Should I seek a patent now, or discover more about what I have? What will that cost? What will the benefits be? What other choices do I have for protecting my idea? Further on in this book, we'll examine the patent procedures more carefully and you'll be able to make some decisions about this subject.

2. Can I sell or license the idea to someone else to make and sell? Most inventors believe this is the route to take. But how do I find a prospect for my particular invention? How can I interest them in considering buying or licensing the idea? What do I have to present to a

prospective buyer or licensee? If you pursue this avenue, here are things you'll need to do:

A. You will need a good working prototype of the product.

B. You should have at least filed a Disclosure Document with the Patent Office and seen a patent attorney about patentability and advice on your legal rights. Your attorney will be prepared to advise you on how to communicate with prospects; you will want to try to do this on a basis of confidentiality. You may approach your attorney to act as a "straw" in your preliminary negotiations; he or she might write the first solicitation letter to the prospect, generally disclose the nature of the invention, and seek a confidential basis for further negotiations.

C. You will have to "go to the literature" and identify prospective buyers or licensors. A prospect would be qualified by (1) having an appropriate manufacturing capability, and (2) having a marketing organization that serves the market into which your invention fits. Check the library as a first stop and look in the reference section for directories of manufacturers. You may also look in the Yellow Pages for local firms and also find there under "Mailing Lists," the names of firms which sell names and addresses. A particularly good source for such information is American Business Lists, Inc., P.O. Box 27347, Omaha, NE 68127, Phone (402) 331-7169. Look also for the *Thomas Register* in your library.

D. You must prepare a complete documentation of your invention. The manufacturer you approach will want to know what the invention is made of, what machinery and processes are involved, what it will cost to make, what the market might be, what the competition is, what it can be sold for, how much sales or marketing expense will be required, and finally, what profit potential there is in the invention for him.

This is a tall order! But rarely will a manufacturer bother to consider a product unless this kind of information is presented. Most manufacturers simply are not interested in ordinary ideas; they get these all the time from their employees and customers. What they're more apt to consider is a "profit center" that fits their existing production and marketing capability.

The information required by a prospective licensee or buyer of your invention is in fact identical to what a venture capital firm will require if you decide to go it on your own and must solicit funding of your venture. So this is basic groundwork you must lay in any event save that of having all the money you need already and having the capability for getting the invention made and marketed yourself. In succeeding chapters, I'll show you how to work up this information, so be prepared for that!

3. Should I make and market the product myself? Do I have the money to do this? Can I raise capital? Do I have the entrepreneurial strengths? If you elect to pursue this course, here are things you must address:

A. You must establish a business. Even if you seriously pursue the licensing route above, you should establish a business; in the former you will be in the business of licensing an invention, in this case, you will be in the business of making and selling a product.

B. You must, particularly if you are to raise capital, work out a business plan. Do it and write it down. Define your mission and goals, and the steps you intend taking to reach them. Prioritize them, and don't make the mistake of thinking you can't do all this without money. This is a necessary part of finding money if you need to.

C. Work up the same cost and profit projections you would have made up for the purpose of licensing the idea. I'll take you through that further on in the book.

D. Study and understand all the manufacturing steps you will be involved in, even if you contract part of these processes out to others. Identify your supply sources and costs. In a further chapter, we'll look more closely at manufacturing and you will understand more of what's needed there.

E. Study and understand the marketplace. What is your competition? How are similar products marketed? Who will you be selling to? How will you do that? Learn all you can about distribution channels such as wholesalers, distributors and chain buying organizations. Consider direct mail; what will your catalog sheet be and what kind of correspondence and pricing information will accompany it?

Marketing is by far the most formidable of all obstacles to inventors! To help with an insight into this major hurdle, we'll address a full chapter to the subject further on in the book.

4. Rather than license the idea or make and sell the product myself, what are the other possibilities? A couple:

A. Consider making an arrangement with a company to manufacture your complete product for you, and you handle just the marketing. You might even be able to find a company who will handle not only the manufacturing but do the shipping, billing and collecting for you too. Such arrangements are not uncommon.

B. Find a marketing organization who will handle all sales and marketing of the product for you, while you simply manufacture the product. This too is not uncommon. But be prepared to bend to the client's wishes on things like packaging, pricing, etc.

Both of the above strategies are far easier to negotiate than the outright sale or licensing of your idea, but in the case of the marketing organizations, be prepared to address the question of exclusivity. A marketing firm which will be putting manpower and money behind your product will very likely want to protect their effort and investment. But this is in fact no less a demand than you would encounter if you were to retain regular sales representative organizations if you were to market yourself through one or more of them. Sales and marketing organizations like the security of protected territories.

We have then, outlined the general courses you might take in commercializing your invention. All of the above are going to require some considerable work; make no mistake about it! We'll explore some of the features of this landscape further on, including a look at what are called the invention helping organizations—inventors clubs, shows, Small Business Development Centers and the like. But you should be clearly aware at the start that there is no structured "system" for getting you to a profitable destiny with your idea. The only system available to you will be the one you develop for yourself, unique to your particular invention and your own special needs. Perish for once and all any notions you may have that inventing is a "get rich quick scheme"!

STARTING UP IN BUSINESS

There is one thing I would urge you to do, however, if you have any serious intention to go forward in any direction with your invention. That is to set yourself up in business. Here's why: It is not expensive to open a special business banking account and to get yourself a letterhead and some calling cards. Then set up a modest bookkeeping system and keep track of whatever expenses you incur in connection with such things as prototyping, patent application and the like. At some point you may buy some other supplies such as catalog sheets, etc., and all of these costs should be calculated. They will be important later on in filing your tax returns.

Formalizing your efforts under the mantle of a business enterprise will also have a psychological effect on you. As the business owner, you will have the responsibility for developing your company's plans and strategies, and in your role as an employee of the business, the responsibility for carrying the plans out. Here are some tips on starting your own business:

Some entrepreneurs prefer starting their own business with a partner. The partnership has certain advantages: the two managers can pool their talents and can share the burdens of decision-making and financial responsibility. However, you should be extremely cautious in choosing a partner. Make no alliances which might impair the integrity of your own performance.

Considering how difficult it is to find a reliable partner, the wisest choice for the inventor is to simply incorporate his venture or operate it as a sole proprietorship. Let's examine the sole proprietorship first.

One of the great things about America is that virtually anyone can go into business for themselves. Last year, 635,000 did! To get the ball rolling, you open a bank account. To do that, you will have to hurdle only a couple of regulatory obstacles.

YOUR COMPANY NAME

If your name is John Jones and you are going to call yourself "John Jones Company," nobody can stop you. If you are going to call it something like "Empire Manufacturing," then you should first register that "trade style" with the Secretary of State of the state in which you live. You may be further required to publish a notice of "assumed name" in the official publication of the town in which you will operate. Consult your attorney on this.

FEDERAL TAX IDENTIFICATION NUMBER

The government will want to know you're in business, for obvious reasons. So when you open your bank account, you will be asked for your Federal Tax I.D. Forms can be obtained from your nearest IRS office, and they can usually advise you on how to obtain forms you will also need for state tax purposes. When these forms are completed and returned, you will be issued a Federal Tax number, and thereafter required to report earnings.

Essentially that's it! You're in business. Of course, your business will be expected to observe all the local, state and federal regulations, just as you as an individual are. You will be accountable for such things as observance of zoning ordinances, minimum wage requirements and local or state sales tax reporting and payment.

CORPORATION Setting yourself up as a corporation is slightly more complicated, but not overwhelming. The procedure is essentially the same except that you must file a Certificate of Incorporation with the state Secretary of State's office. An attorney can put together your whole incorporation package, including your corporate stock registration book, articles of incorporation, bylaws (which may vary from state to state), or in many instances, you can do this yourself.

Pre-printed corporate stock record books are often available at better stationery stores, and these are simply filled in to reflect the requirements of your corporation. If you have a friend who owns a small business corporation, ask to examine his stock record book. The Secretary of State's office in your state capital may also be able to furnish guidelines to assist you in meeting their requirements.

Of course, when in doubt, it is always wise to consult an attorney. Currently, the legal fees for setting up a small corporation run from $600 to $1,000. Matters pertaining to the form of incorporation for purposes of taxation can be discussed with your lawyer, and he or she can provide guidance related to authorized shares of stock. Even though you will be the sole owner, you may wish at the start to have the authority to issue some of the corporate stock to an investor, and therefore, should not limit your authorized shares of stock to just one share.

For those who have not been in business for themselves before, and who are unaware of the process, it can appear somewhat complicated. In fact it is not. Anyone can go into business, including you. Remember only that the incorporation process is always handled by your state, not the federal government. Your only direct connection with the federal government at this start-up stage is the requirement to obtain your business tax identification number. And that is something very much like having to give your social security number when you apply for a job.

SCORE PROGRAM Many states and localities have resources and organizations to assist individuals in setting up their own businesses. The SCORE program, operating in conjunction with the Small Business Administration, can provide guidance. There are also excellent texts on the subject, including some listed in the Appendix of this book. Ask your banker. Talk with other established business people. Find out if there is a local organization offering help. You'll be surprised to find what is available for the asking.

An important thing to remember about this whole process, however, is that these are only the mechanical elements for starting your own business. Like the patent or copyright you may obtain, the corporate stock record book or company checkbook are mere devices. They have their requirements and purposes. But the real heart and soul of your business is you, your self-discipline, your dedication to quality and performance and your personal management skills will count far more strongly. For the most part, you can do it alone. If you need help, try to buy it. If you can't, pick your partners carefully!

Now that we've identified some of the challenges, let's take a look at how to deal with them in the succeeding chapters.

5.

The Art of Prototyping

Many inventions look good on paper but fail to work when you actually construct and test them. Or they don't work in the way you expected them to. So the very first test of your idea's practicality comes with putting one together and trying it out.

Unfortunately, many beginning inventors are poorly informed about materials or manufacturing techniques. For example, people invent things they believe would be nice made of plastic, but they don't know anything about plastics. The inventor must thoroughly understand any medium he wants to work in, whether it be metal, fabric, gear trains, chemical compounds or anything else.

MORE THAN JUST A MODEL Making a prototype is more than making a simple model of the invention. In the prototype, the inventor can identify materials, tooling and fabrication steps to provide information for determining cost and price later. He can determine at the outset the economic feasibility of his project. In fact, the purpose of the prototyping stage is threefold. First, you want to demonstrate functionality. Second, you want to determine the manufacturing requirements. And third, you want a good product example for sales or investment solicitation purposes.

The initial "rough" prototypes do not necessarily have to be made from the material you plan on using for the final product. However, you should keep in mind the benefits and limitations of any of the production materials you plan on using. Part of identifying your production materials is knowing also what tooling will be required. And, finally, of course, you must understand the fabrication or "secondary" steps you'll be using.

For example, an inventor wanted to produce a dental floss pick to fit a certain consumer price category. The object was a throwaway product of polystyrene plastic, with a thread spanning two points on the device. In the prototyping stage, the inventor confronted several tough engineering challenges. Because he wanted the pick to be inexpensive, he had to devise tooling that could produce a large quantity of picks per hour of production time.

The shape and design of the pick itself also presented challenges. The inventor designed a pick shaped as a small bow, but only after much trial and error did he find a way to make the main span of styrene plastic strong enough to hold the string taut. He had to devise a molding process which was both economical and efficient, and which would lend itself to mass production.

The inventor made an original, hand-fashioned sample, so that he could visualize the device. Because he machined the model from plastic, the same material he planned to use for the final product, he could begin to estimate his eventual material cost. The manufacturing techniques were so complex, however, that it was not enough for him to make just a prototype of the pick itself. He also had to construct a "prototype" manufacturing setup too. With the combination of mold and string emplacement in one operation, a simultaneous primary and secondary manufacturing, a fairly complicated manufacturing process became necessary to produce what appears to be a simple device.

In this prototyping stage, many problems surfaced which were totally unexpected. But because of the inventor's persistence and thoroughness, he obtained not only a workable sample of a product, but a definition of the intended manufacturing process too. Obviously, a single item can be made and its intended use tested. But it is equally important to test the manufacturing process.

Even products as sophisticated as the automobile are first prototyped, not only to test the design and function of the product, but also to determine whether existing technology can produce it. And it must be remembered that technology is forever changing.

DEFINING TERMS

People working in prototype development use different terms to define a product sample. Some of the terms are as follows:

Rough: In a 3-dimensional object, an approximation of the design, not necessarily functional or of the specified material, size or color.

In a drawing; a rough sketch, without specific dimensioning indicated.

Comp: In a 3-dimensional object, a precise, functional model depicting all working elements and usually accurate in size, material and finish.

In a drawing, a multi-view presentation of the object including all dimensions, machining specifications (surface finish, edge breaks, tolerances, radii, chamfers and the like.)

Breadboard: A 3-dimensional layout of components or parts to determine the functionality (workability) of a system. Often used in electronic circuit design or to prove out gear ratios, escapements, cams and ratchets in a "train" or assembly of parts.

Stovetop: A lab sample of a formula, as in a food mix, cosmetic, drug or chemical composition. A "stovetop" is intended to illustrate both appropriateness of the formula ratio and the ingredient interaction in processing equivalent to that in actual production.

Prototypes of simple mechanical devices can usually be fashioned of wood or clay. Often, I have used balsa or birch model airplane plywood, available in hobby shops. For clay, I generally use a product known as "Sculpey," a clean plastic material that can be baked in an ordinary oven to a hard, sandable, carvable solid shape. (If not available locally, write: Polyform Products, Inc., 9420 Byron Street, Schiller Park, IL 60176.) I've also found plastic auto body repair filler such as "Bondo" to be an excellent material for prototype construction.

For those doing their own prototyping of simple devices, many of the materials can be readily obtained from local craft, hobby, hardware, lumber, fabric or art supply stores. If machining is required and you do not have a mill or lathe and all of the attendant accessories and skills, find a small machine shop willing to work with you.

PROTOTYPING FIRMS

If your device is complicated and you are not fully skilled in the disciplines relating to your invention, you will probably want to enlist the services of a machinist or fabricator. Regardless of how or where you produce your prototype, always be sure that the quality of workmanship is topnotch and the device is finished appropriately, with good enamel, anodizing or plating, as intended.

There are prototyping firms which do offer good original model making capabilities. Check the Yellow Pages. The better firms will produce drawings, hand machine parts, make temporary rubber tooling and hand cast parts in materials which will simulate the final material. Such firms generally employ a good machinist with knowledge of all conventional machining methods, a sculptor who may fashion a preliminary model in clay, and a fabricator who will design assemblies which lend themselves to the most practical manufacturing steps.

If you decide to commission such a firm to do your prototyping, make sure they have experience working with the materials you want to use for your final product. Ask questions! If your product will be made of plastic, then what plastic do they recommend? Why? Is it a conventional material such as polystyrene, Nylon, ABS (Acetyl-Butyl-Nitrile), or polyethylene? Or is it among the "Engineering" plastics such as glass-filled Noryl (Noryl is a trademark of General Electric) or Teflon? (Nylon and Teflon are trademarks of Dupont.) Can a lower cost styrene be used instead of ABS?

A WORD TO THE TOY INVENTOR

In addition to all the normal considerations outlined here, the toy inventor is confronted with some special ones. Here are several.

1. Toys must pass rigid Consumer Product Safety Standards.
2. The primary objective of a toy is entertainment.
3. The primary purpose of a game is to challenge.
4. What entertains you might not entertain the child; what challenges you might not challenge others.
5. Games should provide quick pace, generous progression and lots of "whoopie" factors.
6. Toys for children must be tested by children. Is your toy chosen first; how long does it command attention?
7. Almost all modern toys and games are variations on those invented centuries ago. There is virtually nothing new under the sun!
8. A key to success is in how contemporary the "new" toy is.
9. Packaging and promotion are critical.
10. If you invent toys, study the marketplace, the industry and the trends.

If your article is to be made partly or entirely of molded metal, problems are similar to those of plastic. A variety of metal alloys, for example, can be molded from a particular tool. Zinc, white metal, pewter and lead all mold in a relatively low temperature range, while copper, bronze and steel require far greater temperatures and different handling techniques. In addition, there is sand casting, investment casting, spin casting, and permanent mold molding.

At our prototype stage, some of the really important questions to be answered concern whether your part can be made better (and less expensively) in plastic or in metal. Likewise, you might con-sider a variety of other materials. Glass, ceramic, cloth, paper and wood are a few alternatives. And can extruded metal or plastic shapes be used, rather than casting and machining components? Obviously, the inventor can make better judgments from a broad base of experience with both materials and processes. But to an extent, you can compensate for a lack of experience by hiring a firm knowledgeable in such matters, and by applying their knowledge to your problem.

So far, we have talked mostly about mechanical devices. But many inventions are chemical or process ones, and the same rules apply to them. Patents are also issued on such things as electronic circuitry and plant varieties.

Notwithstanding, there is a prototype stage to all of these too. A process invention often goes through many "after-steps" before the best technique is found. For example, in one product development, a particular chemical reaction was used to remove calcium from the product. Initially, a sequestrant was used: sulfamic acid. Eventually, in a later "after-step," it was determined that citric acid, a much less expensive chelation agent, achieved the same result. Such is but one example of the usefulness of prototyping in a chemical process area.

Obviously, the inventive process does not end with that first flash of inspiration. It is also the 99 percent perspiration that Thomas Alva Edison spoke of. And the prototyping stage is where a lot of that sweat takes place. It is a necessary step in the journey to success. So, it is important that this groundwork be well laid, even if it involves considerable expense. Prototyping helps you to discover not only whether your "baby will fly," but how well, and at what cost.

The treachery of the prototyping craft is that it is not distinctly defined; it fits no ordinary convention. Inventors invent all manner of things; craftspersons of every type translate these ideas into tangible objects. Yet it remains an elusive and painfully frustrating endeavor.

CONSTRUCTING A MODEL

Given that a worthy invention deserves creditable illustration, inventors absolutely must reduce the invention to practice by constructing a working model. If possible—indeed sooner or later—the working model must have all the functionality and appearance which the ultimate buyer would expect if he were to purchase it in a store. It must be self-proclaiming in appearance and functionally workable. The model must not only demonstrate its usefulness but sell the idea.

Most models, even in disciplined laboratory environments, begin with a kind of makeshift, hopscotch search for materials and components. A scrap of plastic, a piece of wood, some off-the-shelf nuts, bolts, springs, fasteners, bearings, motors, relays, solenoids, air hose, string, nails, clay, fabric and whatever it takes to fashion that first example of the invention.

Professional model-makers are artists extraordinaire. And while the machinist, operating in the realm of tolerances and other engineering requirements, is often employed to produce the mechanism or working parts of the prototype, the sculptor and artist must "package" it and often, as in the case of a telephone instrument, packaging—the outward shape and style of the device—becomes very much a part of the functionality of the completed model.

While the first objective of the prototyping effort should be directed to proving the workability—and superiority—of the invention, an equally important objective is style and purpose. Too many inventors fashion models which are crude, amateurish and fragile. Skilled model-makers, familiar not only with manufacturing processes which dictate to the product's design features, but also with materials which will provide the necessary strength and appearance, often employ simulation processes and substitute materials to arrive at a model.

Model makers often employ silicone rubber (RTV or room temperature curing compounds) to make temporary molds from which cast plastic pieces are derived. The casting resins employed fall into a wide range from epoxy to urethane to acrylic to polyester or even plaster.

These casting resins are selected to imitate, in a general way, the characteristics desired in a production molded piece. Thus, if the invention is ultimately to be produced in a flexible material such as polyethylene or vinyl, the model maker would employ one of the castable urethanes or vinylsols having the same approximate flexibility. If the ultimate part is to be molded in a rigid material such as styrene, then the casting resin might be epoxy, acrylic or polyester.

When preliminary invention parts are hand-cast, frequently the first moldings are crude and often lacking in detail. Then, just as a sculptor "stages" the design of a new coin, the artisan must make further modifications to the first cast part, remake another set of silicone molds, and cast a second-stage part. Sometimes this process is repeated several times until the object assumes its proper characteristic. In these stages, the artist may add engraving, join two or more sub-parts together, introduce cosmetic effects such as surface patterning and features such as internal ribs, insert holes or cutouts, and the like.

Finally, the perfected part is carefully checked for draft, mating with other parts and other important tolerances, both with respect to its own functionality and with regard to the ultimate production molding requirements. This is critically important, because again, many inventors are simply not knowledgeable about conventional manufacturing processes, and thus create designs that are either impossible to manufacture or require such extraordinarily complex tooling that commercial feasibility goes out the window!

EXPLORE THE MARKETPLACE

Inventors also frequently ignore conventionally available parts or components which might be embodied in the invention, and instead, go ahead and reinvent the wheel. The industrial supply marketplace provides a wide range of parts such as motors, nuts, bolts, bearings and the like. Often, a stock extrusion or other part can be employed in both fabrication of the model and in the ultimate production-level device.

While all of the above illustrate particularly the development of a mechanical invention, the same kinds of rules apply to other inventions. For example, a toy inventor is well advised to understand the conventions of say, game board manufacture. Game boards are printed and sometimes die-cut. The manufacturer's sheet press size and die-cutting capabilities should not be ignored. Incorrect sizing could easily add manufacturing costs which make the final product commercially unfeasible. In the same category, game parts such as randomizers or pawns are routinely manufactured by a number of firms, most of whom are eager to work with inventors and can often shortcut the prototyping process while lending coherent direction to the most economical final manufacturing process.

Many small machine shops provide prototyping services, but most have their own specialties and often these do not embrace the full scope of the invention. A machinist specializing in injection mold tooling may have no experience in making rotational or blow molding tools. The invention may also require tinwork, welded framework, ceramic, glass, wood or other materials and the inventor will need to seek out craftspersons skilled in these other arts.

Professional prototypers apply incredible ingenuity in hunting down pieces and parts from which to construct models. A favorite haunt for many is the used machinery dealer or "junk yard." Engineers in Silicone Valley, or Boston's high-tech Route 128, are frequent visitors to such places as Radio Shack or the surplus electronic supply outlets in their areas. One enterprising toolmaker in the Minneapolis area makes frequent tours through several junkyards, with his eye sharpened for someone else's discarded blanks, shapes or parts. Often, components can be purchased for little more than scrap metal pricing, and very closely fit the needs of the model.

Needless to say, many organizations developing new products often define creative environments which shortcut the more tedious formality. In highly structured and formal prototyping environments, the engineer who conceptualizes the product is too often bound to a tedious process of having the idea reduced to articulate drawings by the draftsman, then further "engineered" by department heads, checked by the legal department, etc. Engineers complain that such formality and structure dilutes the vitality of the project, stifling creativity and innovation.

THE "WHAT IF" MENTALITY

3M, Hewlett-Packard, Wang, Xerox, Bell Labs, Western Electric, General Electric and other classically innovative companies all define product development activities, including the important prototyping function, in as loose and stimulating a format as possible. Engineers can quickly avail themselves of a machinist who will translate a rough sketch into a finished part. The "what if" mentality is more than a clever advertising slogan; it is very much a part of the thinking of engineers in such companies.

"What if" frequently leads to borrowing the neck of some bottle, the head of some bolt or knob, a piece of vinyl suede from the fabric store or a section of some antique picture frame. In one instance, a prototyper found a small piece of anthracite coal from which to develop a tooling pattern for a small gold nugget molding.

The object, of course, is get to what is referred to by Tom Peters, coauthor of *In Search of Excellence*, as "the chicken test." The phrase itself had its origins in a classic prototyping experiment. Engineers developing a sophisticated jet engine wondered what would happen if the airplane were to encounter a flock of birds, even geese! Someone suggested, "Turn it on and let's throw a chicken at it!" Plastic or die-cast molders refer to "the disaster shot," that first squirt of molten metal or plastic into a production tool. If the tooling is properly made, and all the design features developed in the prototyping stage are correct, the part will eject effortlessly and perfect from the mold. If not, it's "back to the drawing board," or the toolmaker's bench.

DISASTERS TURNED VICTORIES

It is, of course, to the idea that "disaster shots" needn't be disasters but rather victories that experienced prototypers attempt to work out all the "bugs." Often, an unskilled designer-prototyper creates objects of manufacture which are virtually impossible to make by any conventional means. Production people necessarily concern themselves with "mold flow" within cavities, drafts, sprue runner location, mold cooling necessities and a variety of tolerances necessary for not only proper ejection from the tool, but good finish and dimensional tolerance. Prototyping should in part seek to address these concerns long before production tooling is created.

Notwithstanding the breadboard or preliminary prototyping efforts, the objective is always to get as quickly as possible to a product which looks good and works well. Inevitably, as the prototypes are developed, new insights will develop. Inventions, which are after all unique, have a tantalizing way of leading anyone working with them into an endless round of perfections and more perfections, and the danger is that there becomes no end to the process. Sometimes, product developers spend so much time adding the ultimate feature, that competition beats them to the market!

It is not unusual for the prototyper working with a new product to discover features which the original inventor had not even considered. The prototyper will spot two features which might be more economically combined into one, or a feature which has neither functional nor cosmetic value and would be better eliminated altogether. This leads us to a consideration.

Novice inventors will soon discover that machinists, toolmakers, sculptors and others engaged in prototyping activity are almost without exception a "breed apart." Inventors complain bitterly about slow performance, a negative attitude, a reluctance or downright obstinacy on the part of many of these craftspersons. Call it "artistic temperament" or whatever you wish, but as one prototyper put it, "You bet

I'm obstinate and temperamental, I don't choose to participate in anyone's disaster."

What in fact is often happening is that the craftsperson's mind is racing and the inventor's device is being challenged. The prototyper may hear the inventor specify a certain configuration and disagree. And often the prototyper, more knowledgeable about the ultimate manufacturing method, is right. But there is, however frustrating the process may seem to the inventor, an upside. The prototyper's "devil's advocacy" serves to sharpen and functionalize, perhaps even add valuable benefits to the invention.

One inventor characterizes his dealings with his machinist as roughly equivalent to doing business with a conservative Japanese businessman. "We talk about pigs, chickens, snowstorms, his kids and everything but my invention," he says. "But when he damned well feels like bringing it up, he talks about making my part. What I've come to realize, however, is that all the while he's talking about all this other stuff, his mind is racing. I just have to listen carefully, and eventually he has everything figured out."

For many inventors, this process is burdensome and painfully slow. The inventor is anxious to get his model made; the craftsperson seems to be in reverse gear! One inventor complained, "Yeah, we got the model made, but I had to haul coffee, sandwiches, whiskey and parts into his shop for him, help him fix his broken sink drain and baby-sit his kids to get it done." Obviously, some highly personal relationships develop between inventor and the prototyper, none of it without a certain emotional content.

Unfortunately, prototyping is a process that simply must be undertaken. Until it is done, and the product proven out, there is simply no product; only an idea. Prototyping is a uniquely hands-on process. Edison, in his quest for a filament for the incandescent light bulb, went through numerous experiments before achieving success. That was prototyping in the finest sense. The inventor truly worthy of the definition must likewise become accustomed to this tedious but necessary process.

6.

Calculating Costs and Selling Price

Closely related to prototyping is a realistic cost assessment. Many inventors, caught up in the enthusiasm of inventive "parenting," lose objectivity about costs. First, they are often ignorant about many kinds of production costs. Second, they tend to forget tooling amortization or poorly factor it in.

For example, if you have a plastic mold made that cost $1,000, and the production piece part molding cost is $.17, you should also estimate the number of pieces you expect to sell in a year's time, and then allocate an additional dollar and cents cost to the molded cost to pay for the tooling. You may find that your molded cost of $.17 will double or triple. But it must be realistically calculated in.

Presently, IRS regulations allow a plastic molding tool to be written off entirely within one year. Good business practice dictates that you should probably, but not necessarily, take advantage of this kind of writeoff. Similar IRS rules apply to other kinds of special equipment, but you should check with your accountant or tax person, or even the IRS itself for a specific ruling.

A second problem with cost calculations is estimating labor costs. Many inventors calculate costs based on the assumption that their spouse and children can help out with the project. Such an assumption is often realistic, but you should be sure to add a labor cost factor for family contributions.

Don't forget the value of your own labor. While at the outset, you may not directly pay yourself for various tasks performed, consider what that cost might be if you were reasonably paying someone else to do the task. In fact, any calculations based on the assumption of "free" labor will invariably lead to trouble. With no pay, enthusiasm and performance wane, the quality of the product suffers, and disaster is but a step away.

TRY OUT YOUR COSTS   One of the best methods of fixing cost is to do a small "dry run" on the product. If possible, do a time study. Then calculate a per piece price for each element of work, and offer it to your family, or anyone else willing to do the job. But don't forget, even per piece labor must, in the end, equal minimum wage, either on a state or federal level, depending on which set of rules applies to your product. Generally, if you are selling your product inter-state, you must abide by federal minimum wage laws. If you sell only within the boundaries of your own state, you may have to abide only by state laws. In any event, here again, it is well to consult with your accountant.

DEVELOP COST SCENARIOS   In developing cost calculations, lay out a spreadsheet and set up categories for each potential item of expense (see Appendix). These categories may include, for example, rent, insurance, utilities, raw materials, outside manufacturing charges, assembly labor, packaging materials, supplies, workman's compensation insurance and your portion of the Social Security (FICA) contribution.

When the spreadsheet format is completed, you will then work out several "what if" projections where you will insert estimated costs. Some costs will be in weekly or monthly terms. Other costs will be hourly, such as for labor, if that is how you pay your help, and other costs will be per unit, such as for parts bought outside.

After you have entered all your cost estimates, you must then arrive at an estimate of how many units you can manufacture within a given time frame, perhaps a month. You may want to make several estimates. You will then divide your total dollar expense by the estimated number of parts manufactured and arrive at a per piece or per unit manufactured cost. You may be surprised at how much the cost of your article has increased over the original manufacturing estimate for the major component of your invention. But as in the prototyping stage, it is absolutely necessary to get a realistic picture of where you stand.

Part of the unknowable treachery of the cost calculation process is that you still don't know if the product can be sold. Is there a market for it? Still, you can't begin to answer that question until you first price your product to the trade. How would anyone know if they want to buy one if you can't quote a price? And how can you quote a price without knowing what your costs are? A word of caution: Little accounting mistakes can lead to big problems. For example, transposing units of measure, such as dozens into hundreds or ounces into pounds, can provide misleading results. All units must be translated into equivalent value.

The inventor would be wise to consult an accountant who understands manufacturing cost accounting. Not all accountants are familiar with the special nature of production accounting. Find one who understands the needs you have, and who can also advise you well concerning capital equipment amortization and taxes.

After you've calculated the "what if" scenarios with your spreadsheets, you must arrive at a pricing schedule which will permit you a reasonable profit. Most such price schedules are determined by "marking up" your costs. Generally, you should at least double your cost, and often it can run 3 to 6 times your manufacturing cost. The reason for such a high markup is that you must include your distribution and marketing expenses. Generally the seemingly large markups are based on the need for quite extensive sales, advertising and marketing cost.

Some products, such as food items, are marketed through well established channels. Food brokers, for example, are paid a specific, fixed commission for sales. This fixity makes it much easier to determine a markup. Likewise, manufacturers' reps generally charge a fixed commission and the pricing formulas are similar.

CHANNELS OF DISTRIBUTION   On the other hand, if you must go out with your own salespeople to make personal calls on buyers, you have a greatly added expense. Or, if you add special advertising costs to the effort, the expenses can become formidable. Depending on your product, different channels of distribution utilize different pricing, markup, cash discount, and other terms. You must familiarize yourself with the formula used for the market you are approaching. Go to stores or distributors and ask them what markups and pricing schedules they would need for a product like yours. Most will tell you.

Sometimes pricing practices in an industry will completely surprise you, and even put you out of business. I once approached several gift shops with a new product line of bronze art. I discovered a horrifying pricing formula that literally put me out of business. A giftware wholesaler triples his cost and sells it to the retailer who then doubles the cost! Thus if a product was to retail (to the consumer) for $10.00, I had to sell it to the wholesaler for $1.20. And that price had to include my profit! My raw materials alone cost more than the $1.20 figure, to say nothing of the artist's commission, the mold costs, packaging, labor, overhead and all the rest. The project was abandoned.

ESTABLISHING PROFITABILITY   Now roll up your sleeves because we're going to work through all the costs of manufacturing and selling a product—your invention. When we say selling the product, we don't mean selling the idea to someone, we mean producing and selling the product as a business endeavor. Whether you undertake to manufacture yourself, or do in fact sell the idea to someone else, somewhere along the way, someone has to calculate the costs of manufacturing and selling product. You would if you were to do the manufacturing, a prospective buyer or licensee would have to if they were to form a judgment of your idea.

Some charts are included in this chapter, and if you have access to a copy machine, you may want to make copies for preparing your own worksheets. With these forms, you will construct your costs and be able to arrive at unit pricing for your product. But you will most importantly be able to estimate the potential profitability of your invention in the marketplace.

For those who are not familiar with manufacturing cost accounting, here's some explanation of the process. In every kind of business, there are essentially two kinds of costs. There are (1) direct costs, which are attributable just to the cost of making the product itself, and (2) overhead (or burden) costs which include things like rent, owner's salaries, insurance, and other expenses which go on whether anything is produced or not.

In order to calculate the profitability of a business venture, you must work through several steps. First, you must define your direct costs of manufacturing a single unit of your product. Second, you must determine, as factually as possible, what your monthly overhead costs will be, whether you sell a single item or not. Third, you must estimate how many units can be sold in a month. Then, you must calculate a selling price which will cover both direct and overhead costs and also give you a profit. The two big questions plaguing everyone working through such a profitability scenario are (1) how many can I sell, and (2) can they be sold at the price necessary to cover costs and make a profit? In this chapter, we're going to take a stab at it, but when you get into the next chapter on evaluation, you may uncover some insights which will cause you to want to modify your calculations here.

Use the chart that follows to estimate the equipment you will need to manufacture your product. There are two classes of equipment, basically separated by the way in which they may be depreciated. Equipment wears out, of course, and for tax purposes, you are allowed to write off a portion of your costs as normal expenses.

Major equipment, such as a plastic molding press, a truck or a packaging machine, must be depreciated over a longer period of time, usually three years or more, depending on the formula you apply. Discuss this with your tax accountant. In our example, we're going to use a three year basis, and in that formula, you may deduct 25% the first year, 37% the second, and 38% the third. As you get into this chart, you'll see how this works out on a practical basis for determining your costs.

Tools, dies and fixtures are a category of capital equipment expense which can be written off or expensed out much more quickly; they can be written off within one year's time. So the depreciation expense will be much higher, and must be calculated separately.

SCHEDULE 1
Calculating Fixed Assets & Depreciation

Item	Qty.	Cost	Extension

Machinery & Equipment

Item		Qty.	Cost	Extension
Item 1	_____	_____	_____	_____
Item 2	_____	_____	_____	_____
Item 3	_____	_____	_____	_____
Item 4	_____	_____	_____	_____
Item 5	_____	_____	_____	_____
Item 6	_____	_____	_____	_____
Item 7	_____	_____	_____	_____
Item 8	_____	_____	_____	_____
Item 9	_____	_____	_____	_____
Item 10	_____	_____	_____	_____

Total Cost: _____

25% 1st Year Depreciation: _____

(Divide by 2920)

Hourly Depreciation Factor: _____

Tools, Dies & Fixtures

Item		Qty.	Cost	Extension
Item 1	_____	_____	_____	_____
Item 2	_____	_____	_____	_____
Item 3	_____	_____	_____	_____
Item 4	_____	_____	_____	_____
Item 5	_____	_____	_____	_____
Item 6	_____	_____	_____	_____
Item 7	_____	_____	_____	_____
Item 8	_____	_____	_____	_____
Item 9	_____	_____	_____	_____
Item 10	_____	_____	_____	_____

Total Cost: _____

100% 1st Year Depreciation: _____

(Divide by 2920)

Hourly Depreciation Factor: _____

COMBINED HOURLY DEPRECIATION: _____

The purpose of Schedule 1 is twofold, then. First, it is to estimate the amount of investment in equipment and tools necessary, and second, it is to determine what wear and tear expenses must be considered in calculating your product's cost.

Enter the description and cost of major machinery and equipment in the first section of the table. Add the total cost up. Then, assuming this will be your first year, calculate 25% of that total and enter that amount under depreciation. Then, for purposes of the cost accounting we will do further on, you must calculate what that depreciation amount is for a single hour! The formula I prefer is this: There are 365 days in the year, and assuming a work day of one shift (8 hours), multiply 365 times 8 and you arrive at 2920 depreciation hours in the year.

Divide your depreciation amount by 2920 to arrive at your hourly depreciation figure. You will use this later on in your cost calculations.

List your tools, dies and fixtures under the second section of this schedule. Here, you would list items like small hand tools, forming dies, injection molding molds, assembly fixtures and the like. Your depreciation calculation for these kinds of items will be 100%, and thus the total cost and the first year's depreciation figures will be the same. Then, apply the same 2920 formula used above, and arrive at your hourly depreciation for this category of equipment.

Finally, add up the totals from both categories and enter them in the third section as combined totals. When you've finished Schedule 1, you'll have three main figures. You'll know what your total investment will be in both categories of equipment. You'll know what your total combined depreciation will be for that equipment, and you'll have an hourly depreciation factor for use in your further analyses of costs.

In Schedule 2 you will estimate your fixed monthly costs, sometimes called "overhead" or "burden." As best as possible, enter an amount for each category of expense. As a very rough guide, I've included a typical set of categories for a new small business, but these can vary widely, depending on the nature of your activity. You must determine your own, based on the realities of your location and situation.

When you've entered your amounts, add them up, and as with Schedule 1, divide the total amount by 2920 to arrive at an hourly burden factor. This too will be used in your further calculations. But, you may have noticed, we are beginning to creep up on another important

---

SCHEDULE 2
Calculating Overhead Costs

Fixed Monthly Overhead	Estimate
Advertising	_____
Auto Expense	_____
Bank Charges	_____
Client Relations	_____
Dues & Subscriptions	_____
Electricity & Power	_____
Heating	_____
Insurance	_____
Interest Expense	_____
Leasehold Expense	_____
Licenses & Fees	_____
Miscellaneous Expense	_____
Outside Services (Not Production)	_____
Postage & Freight	_____
Printing & Typography	_____
Rent	_____
Repairs	_____
Salaries	_____
Salary Payroll Taxes	_____
Telephone	_____
Travel	_____
Budget Total:	_____
(Divide by 2920)	
HOURLY BURDEN FACTOR:	_____

piece of information. You've estimated your monthly fixed costs, and from that figure, together with the capital equipment costs and others to come, you are beginning to home in on what kind of capital will be required to start and operate the venture.

In Schedule 3, you will calculate what every raw material component cost is that goes into your invention, how many of each component are needed to make a single unit, and then, calculate the combined material costs for that single unit. For example, it might take three of item 1, two of item 2, and six of item 3 to produce a single unit of your invention. Basically, all we're doing here is asking ourselves how many of what and at what cost each; then we're adding it up to arrive at total material costs.

SCHEDULE 3
Calculating Raw Material Costs

	Component	Qty.	Cost	Total
Item 1	_____	_____	_____	_____
Item 2	_____	_____	_____	_____
Item 3	_____	_____	_____	_____
Item 4	_____	_____	_____	_____
Item 5	_____	_____	_____	_____
Item 6	_____	_____	_____	_____

Total Component Costs: _____
(Divided by Finished Units)

MATERIAL COST PER UNIT: _____

SCHEDULE 4
Direct Labor Cost Calculation

	Hourly Employee	Hours	Rate	Total
Employee 1	_____	_____	_____	_____
Employee 2	_____	_____	_____	_____
Employee 3	_____	_____	_____	_____
Employee 4	_____	_____	_____	_____
Employee 5	_____	_____	_____	_____
Employee 6	_____	_____	_____	_____
Employee 7	_____	_____	_____	_____
Employee 8	_____	_____	_____	_____
Employee 9	_____	_____	_____	_____
Employee 10	_____	_____	_____	_____

Total Payroll: _____
FICA Contributions: _____
Other Contributions: _____
Subtotal: _____
(Divide by Total Hours)

HOURLY LABOR FACTOR: _____

The above Schedule 4 is used to determine what your hourly labor factor is going to be for production. The best way to use this chart is to do an actual "dry run" of production, but you can also make an educated guess if you're not yet in production. Figure out how many people you need to put "on the line" to make the production line work. And then, how much are you going to pay hourly? Enter those figures and then add up your total payroll. As an employer, you will have to make a FICA contribution (check with the IRS or your tax person for the current percentage), and you must multiply that percentage times your total payroll, add it in and arrive at a subtotal amount. You may have to make other contributions such as for workman's compensation insurance, union pension or other purposes and you can calculate that percentage and add the dollar amount under "other." Then, take the total hours worked by all employees combined and divide that into the subtotal to arrive at an hourly labor factor.

There are many ways to estimate these kinds of costs, but essentially, this method provides you a single convenient dollar figure which is your hourly cost of production. Later, you will see how this figure can be used in arriving at further estimates of your product's final cost.

This Schedule 5 is for calculating the cost of services which you may acquire through outside contract manufacturing. You may have part or all of your work done by others, and those costs will have to be considered also. This is basically the same kind of calculating as in the previous schedules: how many at what price, and the total amount.

SCHEDULE 5
Calculating Contract Manufacturing Costs

	Services Purchased	Qty.	Price	Total
Item 1	_____	_____	_____	_____
Item 2	_____	_____	_____	_____
Item 3	_____	_____	_____	_____
Item 4	_____	_____	_____	_____
Item 5	_____	_____	_____	_____
Item 6	_____	_____	_____	_____
	Total Contract Manufacturing Cost:			_____
	(Divide by Units Produced)			
	CONTRACT COST PER UNIT:			_____

SCHEDULE 6
Combined Product Cost Summary

Hours of Production: _____

Variable Factors:

    Combined Depreciation (Sched.1) _____

    Overhead Factor (Sched. 2) _____

    Direct Labor Factor (Sched. 4) _____

Total Hourly Variable Cost: _____

Shift Cost (Sched. 4 Subtotal): _____

Enter Units of Production: _____

(Divide shift cost by units produced)

Per Unit Variable Factor _____

Fixed Factors

    Material Cost Factor (Sched. 3): _____

    Contract Manufacturing Cost Factor (Sched. 5): _____

Combined *Unit* Cost: _____

Case Cost Computation

Units per Case: _____

Shipping Container Costs: _____

Combined *Case* Cost: _____

Schedule 6 above is where you wrap up your actual product cost. By using this method—a little unorthodox—you are able to obtain an estimate of what your product costs to make, which includes all the overhead burdens such as your salary, telephone expense, insurance, etc. With the final figures from this kind of calculation, it is a relatively simple matter to calculate how many units you might have to sell in a given period of time to make a targeted profit.

Some accountants will take exception to this method on the ground that it places the entire burden of your operation on the production side of your business. That is, you may produce goods for only a week out of the month, and yet you are, with this method, saddling all the overhead costs for a month against the production of that single week. The reason I prefer this method is because later on in the marketing of your product, you are going to have some extraordinary expenses such as sales commissions or advertising costs which you did not factor into your original estimate of overhead under Schedule 2. Such is always the case, and it is better to load all your estimated costs into the "bottom end" of your endeavor than to find there is no place to accommodate them later on.

There isn't room in this book to carry the basic calculations you have done in this chapter into all the possible financial projections you might wish to explore. But by constructing conventional spreadsheets such as appear here and in the Appendix, and customizing them to meet your specific needs, with the basic information you have compiled here, you should be able to lay out what are known as pro forma financial projections for your business. This is simply a month by month characterization of what your costs are going to be, how many you expect to sell, and what profit or loss you might arrive at. Over the span of a year, if you control your costs carefully and sell the quantities of your item you hope to sell, your business should profit and thrive.

For those who are familiar with computers, it is recommended that spreadsheets be constructed in one of the popular programs such as Lotus 1 2 3, Twin, Excel, or others. These computer programs permit quick adjustment of figures and allow an unlimited number of "what if" financial scenarios to be quickly calculated.

Again, the purpose of this chapter is to familiarize you with the kinds of information you need to develop, and for more thorough help, you should consult with an accountant who can also advise you on such things as the choices available to you for depreciating capital equipment, what current FICA contributions are and required insurance coverage in effect in your state.

7.

Intellectual Property Protection

Since writing the first edition of *The Inventor's Handbook*, I have uncovered a growing concern among many with the adequacy of our intellectual property laws. At best, the system scarcely protects even large, well-financed companies from the predatory practices of foreign competitors, and it serves the independent inventor with limited resources even less perfectly. I'll explore that in more detail at the end of this chapter, but suffice it to say here only that the system we do have for protecting intellectual property rights is far from perfect. Yet, it is what we have, and if you intend licensing your invention to another company, many will not even consider your proposal unless you have applied for or earned a patent.

My own principal objection to the system is that the patent law has for the most part, no "teeth" in it. There is no criminal penalty for infringing a patent and no fine. The patent merely grants you the right to defend the property defined by the patent. And defending against infringement is a process invariably far too costly for the individual inventor. As many inventors have told me, it is really a game for the "big boys."

Still, the situation is not all bleak! The infringement of a copyright carries a possible ten year prison term and a $10,000 fine, and in that case, the FBI is called in if it isn't too busy with other major debacles!

Nevertheless, let's begin with patents. To get one, you have to tell the government about your invention. If ultimately it results in a patent, then anyone who wants to can find out about your new idea. That's where the paranoia sets in! Many inventors, lacking understanding of how to protect their ideas, become hobbled with a secretive nature. Sometimes, the secrecy has legitimacy. Other times, it is the result of ignorance or ungrounded suspicions. While your disclosure is in the application stage at the patent office, it is kept strictly confidential and you have nothing to fear about confidentiality.

APPLYING FOR A PATENT   When applying for a patent, you may want to hire a good patent attorney. But for many, this added expense is often shockingly high. For the beginning inventor, the costs are sometimes impossible to manage. If money permits, however, there is one important reason for retaining a patent attorney. In a well staffed firm, the assigned attorney will be skilled in the specific category within which your invention fits. If you choose a smaller firm, you should ask specifically if the attorney has practiced in the area relating to your invention. If, for example, your invention is mechanical, he or she should be knowledgeable about mechanical things. Then when you meet with the attorney, you must be prepared to discuss, in full, the nature of your invention.

At this point, the attorney may advise that you do not even have a patentable invention. Yours may be simply a "marketing strategy" or something other than a bona-fide invention. And you cannot invent something which already exists in nature. An invention as defined by the present statutes is a new and novel article, concoction or process which has functionality. Letters Patent cover this kind of invention.

The patent attorney will assess your invention's patentability and generally recommend alternatives if it is not patentable as currently defined. Some of the alternatives are to seek what is known as a "design patent," a "copyright" or a "trademark," any or all of which may form part of your portfolio of protection.

The Patent Office serves this basic essential: in exchange for your full disclosure of your invention, they will give you a patent which essentially grants you the right to exclude others from making, using, and selling your invention. A patent does not pass judgment on the commercial merit of your invention. Nor does it grant you the right to make and sell it. Your invention may be otherwise illegal; for example, it may fail to meet the standards of the Food & Drug Administration. You cannot, of course, obtain a patent on an "Eternal Motion" machine, because all of science has concluded this to be an impossibility.

In my experience, working with good patent attorneys is worthwhile though expensive. The attorneys often lend invaluable contributions in defining the product or process. One attorney used to twist and turn the device, muse and reflect, and then casually say, "Of course you've considered that if you do so and so, you can wind up with such and such?" Eureka! He had added a valuable claim to my invention!

The day when you meet with a patent attorney and discuss your invention is known as your "date of disclosure." Your attorney will open a file and start a record on you. Descriptions, notes and even drawings go into that folder. That file, along with the actual date of disclosure, becomes important if you pursue your course through to a patent.

There is another technique for establishing this date of disclosure. At early stages, many experienced inventors write up their invention and have someone knowledgeable in the field of that invention read their disclosure. Then, in the presence of a notary, sign and have the confidant sign the document attesting to his or her understanding of the invention and its practicality.

THE DISCLOSURE DOCUMENT PROCESS

The Patent Office has a procedure for filing a disclosure document. Your disclosure, in appropriate size and form, is mailed to the Patent Office with a $6.00 registration fee and a self-addressed, stamped return envelope. They will register and stamp your document and return it to you. This will authenticate the date of conception but you will then be required to file a formal patent application within two years from that date, or no further protection is possible. Your idea passes into the "public domain."

The philosophy behind all this "legalese" is well founded. The law assumes that a truly valuable invention is a property which should be protected, just as you would lock the doors to your home or car. If you do not practice what is known as "due diligence," that is, a protective attitude toward your "property," you will have displayed to the world at large and the courts, in particular, a lack of custodial or proprietary conduct concerning your invention.

FILING YOUR OWN PATENT APPLICATION

Thus far, we have discussed the conventional and, for many, most appropriate course for patent application. There is another way, however, and that is to file your own patent application. While this may seem intimidating, with a little examination of the procedures and a proper frame of mind, you can do it! But be forewarned. Writing the application document requires great skill. You could wind up with a patent, but if not well written, particularly with regard to claims, your patent may be weak and perish later in a court test! In fact, nine out of ten patents granted are overturned in the court system.

Let's begin by discussing the major patent classification, that of a Utility Patent. The first step is to obtain, either through your library or from the Government Printing Office, a copy of "General Information Concerning Patents." To order directly, address your request to:

United States Government Printing Office
Superintendent of Documents
Washington, DC 20402

Specify Catalog #003-004-00619-6. Enclose a postal money order for $1.25.

Study this booklet carefully. You will note that the Patent Office has strict and precise rules, and if they are not followed exactly, your application will be rejected out of hand. You must understand that, in a sense, your application for a Utility Patent is a form of legal "pleading." You're willing to disclose what you believe you've invented, they're going to try to prove that you haven't really come up with anything new. But if they can't, and your invention fits the definitions established for a patentable invention, you will be "awarded" a patent.

The format of the document you present and the language you use must cover several essentials:

1. Your name, address and an oath of disclaimer which states that the invention you are presenting is new, is your own personal discovery, and to your knowledge, is unlike any other.

2. A description of the invention, what it does and how. This description may be supported with separately attached drawings "in a manner and form prescribed by the Patent Office." Note that patent drawings must be done in a very specific way; in fact, patent drawing is a virtual art form. You may wish to seek an artist skilled in this for drawings you may require.

3. A further discourse is then made into the "state of the art" domain within which your invention fits, and how your invention "advances the state of the art." In other words, why is what the world has not good enough, and how is your invention going to make it better?

4. Finally, having revealed all, conclude your application with one or more claims. These are the heart of your patent, if granted, and the most important part of your application.

In some cases, the inventor may be aware of other inventions which have similar characteristics. These other characteristics may have to do with physical or utilitarian similarity, for example. Or they may have to do with a process applied to different materials but with different results. Then you should specifically address these issues within the framework of item 3 above. It will be your task to identify the variant or novel departure inherent in your invention and argue that these unique differences define your invention as something truly new and distinctly improved over the existing "art."

THE EXAMINING PROCESS

If you are not aware of inventions similar or even identical to the one for which you are filing an application, you may be reasonably sure the patent examiner will. Ultimately, some or all of your patent claims will be rejected by the Patent Office. For this reason, you may want to search the patent records before filing your application. Because of such a search, you may find someone else has already patented an invention identical to yours, or you may be able to identify the distinct differences between like inventions and yours, and address the problem from the very beginning, with your patent application.

A Patent Office search can be costly, if done through an attorney. You may wish to make your own informal exploration first. Most large metropolitan public libraries have complete volumes of all the patents issued by the government. In some cases, abstracts are also available. More than once, I have spent hours tracking through these volumes or microfiche files and, often, saved myself further expense and agony by discovering someone had long ago obtained a patent on the exact thing I thought I had discovered.

More recently, some libraries have installed a direct access computer modem to the patent office in Washington. It's known as CASSIS and through it, you may be able to make your own search of existing patents. Inquire about it. There are also extensive special databases which provide world-wide information on patents. Again, access is by computer modem, accessible via any home computer. One such service is provided by Dialog Information.

In this "do-it-yourself" patent search, look first at the competition. See if you can find a patent number on the product or package, and start with that. If you think there may be other products also patented which could have a bearing on the "state of the art" in which you're involved, get these patent numbers similarly. Then look them up, make copies of any pertinent documents, and study them carefully. One patent may make reference to other patents, in the same way in which you may be making references (Item 3 above) in your application. If so, look these up too. Searching through patent documents will take you through a maze of others, but the pursuit will be enlightening.

CLAIMS ARE THE HEART OF A PATENT

The writing of claims is a crucial part of your application. Claims fall into three categories: product, process, and product by process. Within each category, both broad and specific claims are drawn. Ideally, you will claim the broadest possible coverage on a product or process almost generically defined. But if the Patent Office grants claims at all, it will seek to grant them as narrowly as possible.

Your application must be signed, properly notarized and sent to the Patent Office along with your filing fee. Fees vary. For individuals and small entities, the basic charge is half that of large companies. But there are additional charges, based on the number of claims in your application. If you are an individual and your basic application contains no more than twenty claims, the fee is $170. For each independent claim over twenty, add $17.00.

The booklet mentioned earlier, "General Information Concerning Patents," provides greater detail about fees, as well as a list of patent research libraries, and should be consulted. Remember also that, later on, if a patent is cleared for issuance, there will be an issuance fee of $280 and patent maintenance fees of $225 to $670 at various intervals of time thereafter.

Upon receiving your application, the Patent Office will check it for correctness of form and notify you of its acceptability or of any flaw in the application. This notification will provide you with an application filing number which will identify it through subsequent stages of examination.

In time, usually months, you will receive the Patent Office's first "action." In it, they will enter objections or approvals on each of the patent claims you have made. Claim rejections are made for a variety of reasons. Your claim may be too similar to a claim granted in a previous patent. More often, the claim may be poorly descriptive or too vague or too broad. For example, "a mechanical device capable of flying in the sky" would be poorly definitive.

After receiving the Patent Office's first action, you have up to six months to file your responses, although the examiner may reduce this to thirty days in individual cases.

Again, the form of your response is carefully prescribed and must be followed carefully. You may argue for retention of a claim as originally presented, but you must provide the examiner with compelling reason, offering specific argument to the objection, point by point. In other instances, you may amend one or more claims with more specific or clarifying language. And you may withdraw claims and even substitute others.

Sometimes, this process takes as long as several years: the patent examiner rebuts your arguments and claims, and you in turn supply amended or counter arguments to his or her objections. Finally, when this process has worked its course, you will either be granted a patent or denied one. The process is based entirely on the merit of your discovery and the quality and persuasiveness of your application.

If your application is ultimately rejected, you do have a right of appeal. The information necessary to make an appeal will be provided you by the Patent Office at the time of rejection. Or, you can obtain further information in the booklet mentioned at the beginning of this chapter.

FOR CURRENT INFORMATION . . .

The Patent and Trademark Office Public Service Center can be contacted at (703) 557-5168.

DESIGN PATENTS

Another major form of patent protection is that of a design patent. Here, a patent is granted on a specific dimensional form and ornamental appearance and any functionality provided by the design is not protected.

Design patents are often used to protect such things as an unusual container, an especially ornate escutcheon or similar article where the unusual design feature is the overwhelming value characteristic. Design patents are not as all-encompassing because they are limited to one precise claim. A design patent is useful, however, because it provides protection in an area not covered by a letters patent or by a trademark or copyright.

The procedure for submitting a design patent application is almost identical to that of a letters patent. But it will be more important to illustrate the uniqueness of the design through your accompanying drawing. The fee for an individual or small entity submitting an application for a design patent is $120. Again, consult your attorney or Patent Office guidelines, concerning procedures.

A patent is the most generally accepted legal claim you can obtain for your invention. Very few companies will even consider negotiating with you if you don't have a patent. Some companies will discuss your idea with you if you have made your patent application, but even then, they will caution you that your invention is valuable only after you have been granted the patent.

TRADEMARKS AND COPYRIGHTS

There are two other main legal devices for protecting your "property." One is a trademark. These are issued by the Patent Office and also by many states. A trademark is quite specific, not only in the wording, but also in the graphic portrayal of the mark, including its color, hatching, and the like.

As a company grows, the trademark often becomes of greater value than the patent or patents. The trademark is what the public recognizes when it buys a company's products. Products, processes, plants and people come and go. But a company's name and identity are preserved by its trademark.

Because a trademark is, specifically, an element of advertising and marketing, and a function of a company's merchandising objectives, it will not be discussed in great depth here. While patents and copyrights "protect" an individual's "intellectual discoveries or creations," trademarks define the public marketing insignia or design.

For more information on the filing of a trademark application, obtain a copy of "General Information Concerning Trademarks" directly from:

U.S. Department of Commerce
Patent & Trademark Office
Washington, DC 20231

The other major protective device is the copyright, issued not by the Patent Office, but by the Library of Congress. Revised in 1976, the copyright laws afford very strong protection for the ideas of authors, artists, publishers and certain other craftspersons.

The Copyright Act generally gives the owner of a copyright the exclusive right to do and to authorize others to do the following:

❑ Reproduce the copyrighted work in copies or phonorecords;

❑ Prepare derivative works based upon the copyrighted work;

❑ Distribute copies or phonorecords of the copyrighted work to the public by sale or other transfer of ownership, or by rental, lease, or lending;

❑ Perform the copyrighted work publicly, in the case of literary, musical, dramatic, and choreographic works, pantomimes, motion picture and other audiovisual works, and,

❑ Display the copyrighted work publicly, in the case of literary, musical, dramatic, and choreographic works, pantomimes, and pictorial, graphic, or sculptural works, including the individual images of a motion picture or other audiovisual work.

Interestingly, the law holds that copyright protection exists for "original works of authorship" when they become fixed in a tangible form of expression. The fixation does not need to be directly perceptible, so long as it may be communicated with the aid of a machine or device.

Consequently, the copyright law in a substantial way now protects some of our most important new technology—computer programming! In essence, the assembly of digital electronic pulses, directing either home computers or complex industrial processes, is much like a dramatic script instructing actors and others in the performance of a play.

If you're a software developer, both copyrights and patents will be employed to protect your work. Patents are granted in this area on such things as program algorithms, display presentations or arrangements, menu arrangements, editing functions, compiling techniques, program language translation methods, and operating system techniques to mention a few. Design patents are sometimes secured on a computer screen portrayal and finally, copyrights are employed to protect software against duplication, preparation of derivative versions and distribution to the public by sale, rental or otherwise.

Our founding fathers wisely abandoned the distinction between science and useful arts, authors and inventors. Article I, Section 8 of the Constitution of the United States reads, "Congress shall have power to promote the progress of science and useful arts by securing for limited times to authors and inventors the exclusive right to their respective writings and discoveries."

There are categories of materials not generally eligible for statutory copyright protection. These include:

❑ Works that have not been fixed in a tangible form of expression.

❑ Titles, names, short phrases, and slogans; familiar symbols or designs; mere variations of typographic ornamentation, lettering, or coloring; mere listings of ingredients or contents.

❑ Ideas, procedures, methods, systems, processes, concepts, principles, discoveries, or devices, as distinguished from a description, explanation, or illustration.

COPYRIGHT FORMS The fee for registering a copyright is $10.00, and filing is done on one of the forms available on request from the Copyright Office. The main forms are:

Form TX: For published and unpublished non-dramatic literary works.

Form SE: For serials, works issued or intended to be issued in successive parts bearing numerical or chronological designations and intended to be continued indefinitely (magazines, etc).

Form PA: For published and unpublished works of the performing arts (musical and dramatic works, motion pictures and audiovisual works).

Form VA: For published and unpublished works of the visual arts (pictorial, graphic, and sculptural works).

Form SR: For published and unpublished sound recordings.

Incidentally, a computer program or computer data base would be registered on Form TX. If you wish to order forms only by phone, the Copyright Office maintains a hotline: (202) 287-9100 at any time, day or night. If you have other questions or need assistance, call (202) 287-8700 weekdays between 8:30 a.m. and 5:00 p.m. Eastern time.

COPYRIGHT CIRCULARS For more detailed information, you may wish to order the following circulars directly from the Copyright Office:

Circular R1 - Copyright Basics.
Circular R1c - Copyright Registration Procedures.
Circular R2 - Publications on Copyright.

With Congress' enactment of the 1976 copyright law, a significant strengthening of intellectual property rights took place. Among other things, the law provided that ownership of the created work commenced with its date of conception, with or without notification at the time of publication. It was a landmark overhaul, long overdue.

What was important here was that ownership of the intellectual property became automatic at conception and moreover, an intellectual creation was deemed to have the same legal tangibility as any other property; thus it was subject to the same legal protection as household contents or land. The painting, sculpture, computer program or script, for example, was not merely an illusive intellectual expression, wispy and evanescent, but tangible, ownable and protectable.

But how protectable are these rights? According to Senator Alphonse M. D'Amato of New York, the sale of copyrighted works accounts for more than five percent of our gross national product; in 1985 such sales contributed a significant $170 billion to the gross national product. Yet, according to the United States International Trade Commission (ITC), America's trading partners are marketing about $8 billion of illegally reproduced American intellectual property every year.

The stakes can be high. After committing twenty years and over $200 million to the development of optical wave guide technology, the Corning Glass Works has been unable to prevent Japan's Sumitomo Company from infringing on their patents and exporting the identical technology to the U.S. And this is but one of thousands of such examples!

We hear frequently of the large scale piracy of corporate technology, but what about the independent inventor or artisan? If patented, trademarked or copyrighted goods are so flagrantly stolen, both here and abroad, what incentive has the individual for securing formal constitutional protection which is unenforceable? Moreover, the individual is seldom equipped financially to sustain a battle for his or her rights. While the giant Allied Corporation finally obtained an exclusion order from the ITC against eleven foreign infringers on its patented amorphous metal alloys technology, it spent over $3 million in legal fees, and to date has obtained no monetary damages whatever from its efforts. Equally egregious was the battle Texas Instruments waged for over thirty years with Japanese companies, over its intellectual property rights to microprocessor technology. If the larger company has ineffectual or insufficient remedies, even with extensive cash and legal resources, what chances does the independent have?

THE NEWMAN CASE   Then there is the case of inventor Joseph Newman of Lucedale, Mississippi who filed a startling patent application on March 22, 1979 for an "Energy Machine" which he claims can harvest useful energy from forces within an electromagnetic field—a conversion of mass (matter) into energy. Since energy output was held to be on the order of twenty-five times the input, his invention was categorized as "smacking of perpetual motion" and therefore not allowable under patent office rules.

Notwithstanding, Newman contends his discovery would introduce limitless, pollution-free, and inexpensive power for virtually every need and ". . .will change the world drastically for the good of humanity, more than any invention before this time." The claims would seem incredible were it not that no less than thirty reputable experts—physicists, nuclear engineers, electrical engineers and others—have tested and affirmed the credibility of Newman's claims. And there have been six special bills before the United States Congress calling for the issuance of a "Congressional Patent" to Joe Newman.

Joseph Newman is in a "catch-22" situation. If he has a new discovery, he is entitled to profit from his intellectual contribution. If the Patent Office refuses to legally confirm that his discovery is "novel" (i.e. new and original), non-obvious to a person skilled in the art to which the invention pertains, and useful (i.e. "works and has purpose"), then he cannot market his device without becoming fair game for piracy. It is in the latter category that contention exists between Newman and the PTO. Newman has waged a considerable battle to secure his rights, including a massive public relations campaign, talk-show appearances, and public demonstrations, all supported by some significant outside investment. And there's the key: Without money, his battle would be lost at the outset.

Wilbur and Orville Wright waged a similar three-year battle in 1903 when the Patent Office rejected their claims for a flying machine on the grounds the airplane was "a device that is inoperative or incapable of performing its intended function." Robert Goddard fought for years to validate his discoveries in rocket propulsion; he was awarded tribute and recognition only after his death. Laser inventor Gordon Gould finally won a 25-year battle over his rights to technology which makes present recording discs possible, but it was a tedious battle! Slizard and Fermi found it necessary to sue the United States for monetary compensation when their atomic inventions were simply appropriated by the government in a "right of eminent domain" action deemed necessary to "national security." And a battle is surely shaping up over the various claimants to discovery of "Cold Fusion."

WHERE DID IT ALL BEGIN?

These are but minimum examples—intellectual property right weakness both before and after the legal process. Why? In part, it may lie in the fact that our fundamental constitutional law derives out of old-world legal philosophy originally dedicated to the extension of imperial domains by one crown or the other. For example, the Hudson Bay Company, the East India Company and the Jamestown Colony (as well as others) were all granted as patents. In effect, but with variation, the crown granted "exclusive commercial opportunity" to individuals in exchange for the colonization (or conquest) of other lands and territories.

The reason for such co-ventures was that the patent-holder derived a commercial opportunity while the state extended its dominion and thus presumable power. There is precious little difference between 16th century piracy of colonial goods of commerce and 20th century piracy of our intellectual property. Yet the patent was a remarkable political-commercial instrument. And America's founding fathers incorporated it into Article I, Section 8 of the Constitution by investing Congress with the authority to "promote the progress of science and useful arts, by securing for limited times to authors and inventors the exclusive right to their respective writings and discoveries." Thus was paved the way for the Patent & Trademark Office and Title 35 of the U.S. Code.

There remains an incongruity in the system, however. It is in the phrase "useful arts" as further defined to mean being workable and having purpose. It would be sobering to discover a patent examiner of any mettle who hadn't more than once pondered just how useful a particular invention really was. This suggests the law places an inordinate burden on an examiner to even contemplate the usefulness of what he or she is examining.

Decisions about usefulness are extremely subjective, as any garage sale attendee fully realizes. That is, if you believe in the wisdom of the pocketbook! And that's the heart of the matter. Invention is the very cornerstone of the free enterprise system which is based in it's purest form on the wisdom of the marketplace. Why are patent examiners placed in the incongruous position of passing judgment on a product's merit; it is exclusively to the marketplace we should look for judgment.

Present patent law gives the examiner a decision-making power almost as autocratic as that of a 16th century sovereign. The paradox is that in America at least, sovereignty lies squarely with "the people." True, the examiner works in the public trust, on behalf of the people, but it is surely a convoluted test of representative doctrine.

What if the Wright brothers machine didn't fly? It might have then become simply an objet d'art. Icarus II, perhaps! And as such, it could be quite well protected with the simple filing of a Class VA copyright registration with the Copyright Office. Same thing is true of Mr. Newman's energy machine. In any case, value and usefulness are factors unrelated to authorship; they are perceptions based on judgments of the marketplace.

THE MARKETPLACE SHOULD JUDGE

In argument to the granting of patents without regard to usefulness or even workability is the political philosophy that government must protect people from fraud and deception. A curious reasoning, in view of the fact that if we're competent enough to elect our own representatives, we should be competent enough to make other value judgments. Or do they know something we don't? Moreover, protecting people from themselves is a selective endeavor; witness the limited concern with consumer credit.

Notwithstanding, ample law does prevail elsewhere covering the "caveat emptor" bogey, adequately administered by the Federal Trade Commission and others. Is it necessary to join the patent examination process with any requirement for value judgment whatever?

Only gradually have we come to acknowledge that intellectual property might just have a recognizable quality that is neither associated with commercial profit or nuts and bolts tangibility. Something a medieval artist, secure in a feudal structure, may have sensed better in fashioning a tapestry in tribute to God, nature or "mere transient beauty waxing!"

In a greater context, modern societies place great emphasis on individual rights. Massive changes have and are taking place globally to insure that not only personal liberties but personal property rights not be denied. And there is ample reason, beyond a simple fairness doctrine, to assure that creativity in whatever form be acknowledged and protected. Because in the final analyses, it is from ideas, clever invention and inspiring art that the true prosperity and endurance of every culture derives.

If you are confused about how best to protect your ideas, whether by patent, copyright or simply keeping your secret to yourself, you should consult a patent attorney. But I would also caution against getting overly enthusiastic about how well protected you might really be. Many inventors make the assumption that with a patent or copyright, they have it made. The battle may be just beginning. There are unscrupulous people who will copy your ideas and force you to test them in court. This can become a formidable battleground and meagerly financed individuals often lose.

In the final analysis, as strongly urged before, the quality of your product and service is most important for success in the marketplace. No legal artifice can ever substitute for your continual dedication to that principle of quality!

8.
Evaluation and Feasibility

Determining the success probability for a new product is about as tough as picking the winning numbers in a ten-number lottery! The variables are so extremely complex, it defies almost all rational methodology. Largely, this is because we are dealing with the most complex of all subjects—human behavior. What people will buy or not is always a matter of great speculation. Even seasoned manufacturers bringing a new product to the market find themselves as anxious as a playwright at his first Broadway opening, or a new father in the waiting room.

After months—even years of planning by the manufacturer—the final judgment is passed by the consumer. What a marvelous system! But it is not a perfect one. For one thing, the consumer is too often manipulated by massive promotions appealing not to reason, common sense, or perhaps even truth, and the "wisdom of the pocketbook" becomes muddled or mindless. What we eat and drink, what sports we enjoy, what fashions we wear, and what lifestyles we live are concertedly dictated to us by gigantic promotional and public relations activities.

Yet is this all bad? True, it can be argued that we are being manipulated, but there is another side to the coin. When a company invests millions in a product and all the production facilities that may underlie it, that company is gearing up for mass production. Mass production brings lower cost goods to the public but it is also dependent on mass sales. In order for the manufacturer to be able to invest millions in a product to keep its price low and quality uniform, it is obliged to protect itself by pulling out all the stops to assure consumer acceptance. Thus the enormous promotions. Without them, many products could not be mass produced because there would be only meager public awareness and acceptance.

But long before a manufacturer gets to the manufacturing and promotion stage, there's always a hefty examination of the new product's likelihood for success. In the case of the large, established manufacturer, however, many of the evaluation factors are very well known because the manufacturer knows the extent of such things as his own financial, personnel, manufacturing and marketing strengths. Moreover, he has probably dealt with other products and has historical data as a resource for examining the potential of the new product being considered.

But how does an independent inventor, without any of the resources of an established company, evaluate his or her invention? Tough as that may seem, it's done in just about the same way the large company does it. Tediously! But I must say that while I will try to review for you some of the elements that must be examined to make an assessment of a product's chances for success, it would be arrogant for me to suggest that I can turn you into a feasibility expert in one small chapter in a single book. Understanding what makes a product successful is simply beyond novice comprehension. It is a field of enormous complexity requiring years of study and even then, experts are constantly fooled by the public. Nevertheless, let's start with a review and explanation of criteria often employed, with variation, by professionals examining new product feasibility:

ELEMENTS OF THE EVALUATION

Legality. Can the product be made and sold legally? For example, if a food, cosmetic, medical device or drug product, does it meet Food and Drug Administration requirements? Does it meet the standards of the Consumer Product Safety Commission? Will the conditions of manufacture conform to OSHA guidelines? Are you sure the product (or the process of manufacture) does not infringe on someone else's intellectual property rights? Does the product conform to all local, state and federal laws; for example if manufactured by employed people, are necessary employee records kept and wage deductions made?

Safety. Is the product safe for intended use? Does the product have an "inviting nuisance" quality? That is, would it attract children, unintended or unqualified users to employ it at risk to their own safety? Are there unexposed risks, such as lead content paint, or tendency of the product to fail under circumstances which might cause injury or death? Are warning labels required? Have you obtained insurance coverage?

Environmental Impact. Concern with this issue is mounting steadily. Is the product and its packaging recyclable? Is it biodegradable? Does it utilize materials of manufacture which are hazardous to land, air or water quality? Can hazardous components be eliminated or substitutes made?

Societal Impact. Are people ready to accept such a product? Would it offend contemporary notions, long accepted practices or prevailing conventions? Would the product generally provide a utility or value for "the good of mankind," or does it carry the potential for harm or destruction to our life styles and culture? Aspirin would be an example of a product positively impacting society while certain other drugs could be considered negative in their impact.

Potential Market. Is there a large market for this product? Is it identifiable? Is it reachable? At what cost?

Product Life Cycle. Is this a fad or a product directly or indirectly serving long-term basic needs? Is pricing being structured accordingly?

Usage Learning. Are benefits of the product quickly obvious and can use of the product be obtained without extensive training? For example, the usage learning factor discourages many from becoming interested in computers. Today's children will not, unfortunately, become interested in toys or games which require a lot of learning. If usage is complex, will adequate instructions for the product be provided and will promotion address this concern and dispel it?

Product Visibility. Is attractive packaging required and is it being provided? Will the product's visibility be high at the point of sale? Have all options for increasing visibility, such as negotiating special shelf position or compatible specie location? An example might be that of locating salad dressing in the produce section of a store among the lettuce heads, or merchandising automobile wheel covers with tires, etc.

It should be noted here that retail stores employ every strategy possible to increase the shopper's awareness of merchandise. For example, toothpaste is generally located in the back of the drug store while candy or greeting cards are displayed to the front. Why? Because toothpaste is a necessity and the shopper has predetermined the need, perhaps even goes to the store especially for it. Placing the toothpaste in the back of the store encourages the shopper to walk by many other displays and thus, perhaps, buy something else on impulse.

Candy, cosmetics, jewelry and greeting cards, on the other hand, tend to be purchased on impulse and placed at the front of the store, in view of passersby. These items invite people in. The science of store display and merchandising is a high art form indeed

Service. Does the product require service? Have you provided for that? Must you train users, dealers, salespeople or others in the use of the product? For example, the manufacturers of operating room utensils may actually go into surgical operating rooms and stand by to guide surgeons in the use of new implements. Accompanying service, are their explicit manuals?

Durability. Have you considered all factors which would affect durability? Your manufacturing assembly, component materials, user abuse? Are there any elements within the product which are greatly more fragile than the whole? (Consider the *Challenger* disaster!)

Competition. What is this product up against? What else might be under development "out there" which might adversely affect your success? Will your product stimulate others to develop a competitive product or imitate yours? Do others address the same needs with a better method? Do others address the same needs with less expense?

Product Functionality. Has the product been reduced to its simplest form? Are parts reduced to their minimum number? Is assembly reduced to a minimum? Does the product function as intended, and better than present methods?

Product Feasibility. Is the process of manufacture reduced to its simplest form? Does product design permit "off the shelf" machinery and tooling, or is custom equipment necessary? Is the process of manufacture a proven one? With this product? Are materials of construction readily available through normal supply channels? Are bottlenecks eliminated from production steps? Are all supply sources assured and backup sources available? Are source costs assured and stable?

Development Status. Is the product fully developed? Are all commercialization strategies developed? To what extent is production and marketing in place? That is, is machinery acquired, has production commenced, is packaging done and is final product available for sale? Are catalog and price sheets designed and printed? Are customer prospects identified? Are selling means defined? That is, direct mail, phone, commission salespersons, or by employee salespeople? Are all legal, insurance and other matters identified and requirements met?

Investment Costs. Have you identified the amount of investment, not only for machinery, raw materials and labor to manufacture, but also for marketing, financing of accounts receivable, overhead, and other long-term needs?

Trend of Demand. Do you anticipate that demand for this product will grow or diminish? Does the product fit into an industry which is growing or declining? Are the investment, production facility, pricing and payback commensurate with these factors?

Product Line Potential. Is your product the first of a potential line of similar products? Consider breadth of line: many lengths of 1/2" pipe nipples. Consider depth of line: all kinds of pipe fittings in all sizes.

Resource Potential. Will the equipment, supply, personnel and marketing resources you develop for this product also serve other profit-making opportunity beyond the intended item of manufacture? Can you make and market other products with the same resources?

Need. Does the consumer recognize a need for this product or will they have to become sold on that? How strong and universal is that need? Does the need fall within the category of "basic human needs," such as food, clothing and shelter, or does it fit an incidental or transient need? To what extent will consumers truly need the product, and to what degree will they want it? We are more apt to buy things we want, then things we need! And then we convince ourselves that we truly needed it.

Promotion. How much, what kind and how expensive will be the promotion required for this product? What kind of competitive promotion will you be up against in gaining attention for this product? Caution: Do not expect the novelty or cleverness of your product to eliminate promotional need. Are promotional expenses included in the product's cost?

Appearance. Does the product look good? Is it attractively packaged?

Price. Can the product be priced competitively? Can you make money at that price? Have you allowed for all distribution discounts such as dealer's, sales commissions, wholesaler's, etc.?

Perceived Value. Would the consumer consider the cost of the product fair for what's being purchased? Does the function and usefulness of the product and the purpose of its use justify the price?

Protection. Were you able to obtain Intellectual Property Protection?

Payback Period. Considering the investment and time required, and based on conservative sales estimates, will there be a return of the original investment in one year? Two years? Three years? How many?

Profitability. Is the amount of profit likely from this endeavor equal to or greater than profit which might be obtained from other investments?

Marketing Research. How carefully have you determined that a market actually exists for this product? How well have you identified the consumer and his or her needs?

Research and Development. Have you a continuing research and development program intended to improve the existing product or yield others?

Stability of Demand. Is the product seasonal in nature? Is it greatly subject to sales decline with economic downturn, fashion changes, public whimsy, advancing competitive or alternate technology, etc.?

User Compatibility. Does the product fit easily into existing user formats? For example, if a computer program, would it use existing disk operating systems, if an appliance, existing power supplies, etc.?

Product Interdependence. To what degree is your product dependent on someone else's? To what extent is the product dependent on user skill? To what extent is the product dependent on weather, political or societal caprices?

Distribution. Is the product easily shippable? Are there identifiable channels of distribution to the ultimate consumer? Is freight a compelling cost factor? Is bad weather?

Existing Competition. How formidable is your competition? If there is no competition, how do you know there's a market? Have you a plan for competing with other marketers, other than with the novelty of your product?

Quality. Despite whatever novelty your product has, are you firmly dedicated to quality, service and value? Have you not only a policy but a program for assuring that?

Potential Sales. Have you carefully and conservatively estimated the sales potential for your product, and the costs for obtaining this? Have you allowed a safety factor in those estimates which will permit you to survive with less sales? Is the sales potential great enough to support a company as you envision it? Have you actually made any sales? Have you received favorable feedback? Have you obtained repeat sales? Are customers satisfied?

These, then, are the major factors to look at in judging the feasibility of your product. There are no absolutely clear-cut answers. But there is one more major and in fact, most important factor and that is—you! How can we grade your "feasibility"?

ENTREPRENEURIAL QUALITIES

Your chances for success depend on how entrepreneurial you are. That is a subject too large for a single chapter or even book. But here are a few qualities most everyone agrees are essential:

1. You are a hard worker, dedicated to accomplishment. You do not easily accept "no" for an answer. You are goal-oriented and have an ability to prioritize your efforts efficiently. You act, rather than talk, and do rather than just think about doing.

2. You have good communication skills. You make your needs known to associates and employees and are able to lay down clear directives. You listen for feedback and honor the viewpoints of others.

3. You provide good value to your customers and stand behind your product. You provide good service on time.

4. You value your commitments and keep your promises.

5. You are honest with people and particularly with yourself!

6. You know your merchandise thoroughly and constantly try to improve upon it.

7. You manage your enterprise carefully; you budget your money and resources wisely.

8. You recognize the needs of others—employees, stockholders, suppliers, customers—and serve them fairly and with value.

These might be elaborated upon and surely there are others. Most importantly, if you have a positive attitude, are willing to work hard, and have the determination to face the challenges, you can succeed. And if your product looks good in terms of the factors outlined above, you and the product will make a winning combination!

WHY DO NEW PRODUCTS FAIL?

Most new products never even get to market. Those that do are often poorly managed. It is doubtful that an inventor who does not possess entrepreneurial skill will succeed alone. And what is entrepreneurial skill? Or is that the correct term for it?

More than anything, entrepreneurialism is an attitude. It centers on a desire to provide a valuable product or service which will attract buyers and afford a profit for the endeavor. Ideally, the product or service will be of greater value and lower cost to the buyer than competitive offerings.

Unfortunately, innovation and invention too often focus on a singular mission of reducing factory cost and fail to pass on or share the savings with the consumer.

Many inventors lack persistence. For a product to become accepted in the marketplace, persistence is an absolute requirement. That's why large companies repeat and repeat their advertising messages. That's why salesmen call and call again on prospective buyers. It takes persistence.

But above all, there is no substitute, including a novel idea, for constant attention to all the details of quality, price and service to the consumer. Entrepreneurs who forget for a moment who's boss, go broke.

9.

The Fork in the Road

By now, you have a pretty good idea of what your invention really is. You've probably prototyped it and know that it works the way you expect it to. You've carefully examined what it will take to make and market the invention as a finished product. And you've evaluated its potential for success "out there in the real world." If you've done all this carefully and honestly, you've done the necessary things that most inventors simply fail to do.

Having done these things, you've certainly granted yourself a measure of objectivity that will permit some reasonable judgments about where you want to go with your invention. You may have discovered flaws and weaknesses, but perhaps also some remedies for correcting those. For example, by looking at marketplaces and pinpointing specific prospects, you have some idea of what steps you must take to reach them and what costs would be involved. At the very least—and this is important—you've probably been able to uncover some of the spooky hazards that you might never have guessed lurk ready to scuttle your hopes for success.

TIME TO FISH OR CUT BAIT! But it is also a time for soul-searching. And surprising as this may seem, the most elusive factor in the deciding process is always you, the inventor! Because what you do with your idea is vastly more important than the idea itself. And I want to emphasize again the word do! Not talking or thinking about what you're going to do, but what you actually choose to do.

First of all, you don't have to do anything. If you've obtained a patent on your invention, you've received a pretty patent certificate and you can hang that up on the wall and bore all the rest of us for the balance of your life about your creative prowess! But maybe inventing is kind of a hobby for you and you'd just like to go on and invent some other things. You really don't care about whether anyone else would ever buy your invention or not. Like collecting pretty rocks or better still, doing your physical fitness workouts—all of it is something just for you, and it is not important to your ego if anyone else thinks it important or not. In other words, you're an inventor just for the fun of it. There's nothing wrong with that!

There's nothing wrong with that if that's the truth. Mind if we talk about it? I'm not disputing your word, understand, but I'm perplexed. Are you telling me that you don't really care if other people like your invention or not? Do you mean to say that it wouldn't be nice for someone to look at it and really get excited about your discovery?

Wouldn't you make one for me if I was willing to pay you for it? Are you really sure you don't want to fuss with your "gadget" anymore? After you've done all that creative work? I'm disappointed, but I'll accept that's really your decision. But do us all a favor, then! Don't bring the subject of your invention up again—and again—and again—for the next thirty years! Don't show up at the next inventor's show with the same invention looking for another round of applause. Don't keep coming to the inventor's club and asking the rest of us to analyze the darned thing one more time.

O.K., O.K! Maybe a little public recognition wouldn't be so bad after all. Now would it? Say, we're all human, and it's nice to get a little slap on the back once in awhile, isn't it? Want to know what's even nicer? Having someone write you out a check and pay you for something you plucked from some mysterious corner of the universe—an invention! It's almost as nice as a warm hug! Think about it. But then think about this also:

TO BE OR NOT TO BE

We humans have many marvelous blessings, one of which is that occasionally we are permitted to glimpse discoveries of unimaginable delight. And when someone cares enough to do something about them, they become transfixed in our art and in our invention. We would not have the stunning Rodin sculptures if he had not chiseled their beauty from raw stone. We wouldn't have airplanes if the Wright brothers and others hadn't taken the risks—and done something. We won't have your discovery unless you do something. That's the key. As Shakespeare said, "To be or not to be!"

A large part of the purpose of *The Inventor's Handbook* is to help you make your decision regarding whether you want "to be or not to be." And that is your first choice—to do something, or nothing. So on reflection, maybe you'd like to do something. What?

If you decide to do something, there are two paths from this fork in the road. One is to license or sell your invention, in its present state, to somebody else. The other is to go into the business of commercializing your invention yourself. In either event, if you've gone through the various steps outlined earlier in this book, you have a fairly well-founded basis for making sensible choices and planning meaningful strategies. If you hadn't gone through those exercises of prototyping, financial analyses, legal protection and feasibility, you would be far less prepared to make rational choices. Let's look at outright sale or licensing first.

FINDING A PROSPECT

Most inventors believe that what they'd like to do with their inventions is sell them outright or license them to someone else for commercialization. The big mistake these inventors often make is that they don't properly organize to do it. As discussed earlier in this book, established companies are not looking for just ideas; they want "profit centers." What is a profit center? A profit center is a particular activity within a company which can utilize existing facilities, equipment, personnel and other resources to generate additional profits. The first trick is to find a manufacturer who has the resources needed by your product to fulfill that objective.

Consider this scenario: Suppose you could walk into a manufacturer/merchandiser and say: "Mr. Jones, I see you make and sell widgets to the department store trade and to do that, you must have metal forming, plastic molding, fabricating, and packaging resources. I have a widget that would sell to the same department stores—your sales force could also be selling it—and you have all the manufacturing capability for making it. I'd like to sell or license my widget to someone like you. I've worked up all the financial data. Can we discuss it?"

Suppose you could carry that a step further: "Mr. Jones, I've also developed the special tooling and fixtures that are necessary for this particular product; they've been production tested and our costs nailed down." You bet! That's better still. But let's go another step:

"Mr. Jones, we've also done a market study including a 'focus group' examination, and determined there is a very large market for this widget." And now for the coup de grace: "Mr. Jones, we've already put this on the market and are generating very profitable sales!"

If you were Mr. Jones, would you be impressed with some of these statements? Of course! But note that in each successive stage, you've addressed in ever more compelling fashion, the question of risk for him and his company. In short, the further along you carry your invention, the more of those risk factors you, yourself, address, the more compelling will be your solicitation of a manufacturer to buy or license your invention.

While "new" has its own glamor and excitement, it always carries that element of risk. Perhaps that's why "new" excites us so much; we never quite know what surprises lurk in the newness. But "new" is wrought with dangers too; the public is fickle and the fates are capricious—your product might well carry the seeds for its own self-destruction! Consider the Dalkon birth control shield or the Edsel automobile.

YOU MAY BE THE OBSTACLE

And what about you, as a risk factor? What's it going to be like dealing with you? Can he count on your cooperation to get your invention off and running? Are you a reasonable person to deal with? Are you willing to become a "partner" in a common cause, or are you going to adversarially nitpick the process to death?

All of these, then, are front-end strategy considerations for selling or licensing your invention to another company. But there are others. Who you are going to find to talk to is one. Can you find someone yourself, or must you hire an agent or "marketing" firm to do that for you? I think you'd better do it yourself. It isn't that difficult.

Now go back to Chapter 4 of this book and re-read the section on finding licensing prospects. You're going to have to find them and the only way to do that, short of hiring some rather expensive business consulting, is to dig the information out of directories and mailing lists wherever you can find them. Fortunately, there's a wealth of business directories in many public libraries and you can compile all kinds of lists of prospects. Many directories provide the names of key personnel and even phone numbers. Take along a notepad and pencil and start making your list.

WRITE A PROPOSAL

Second, as also suggested in Chapter 4, you're going to have to prepare a complete dossier on your invention. This proposal should include at least the following:

❏ Your name and address and those of any associates participating in the product's ownership.

❏ The name of the product (invention) and its purpose.

❏ The intellectual property rights status.

❏ The current status of your product development.

❏ The product's features and benefits.

❏ The marketplace(s) it would serve.

Complete cost and pricing analyses including capital equipment requirements, special tooling or fixturing, estimated marketing and administrative costs and a pro forma profit and loss projection. This is the information you've presumably already worked out in Chapter 6.

❏ Any manufacturing or marketing activity accomplished.

❏ Your offer and price!

Outright sale only?

For cash only? Part cash and part stock? How much?

❏ License?

Exclusive? With minimum compensation guarantees? For cash only, or part cash, part stock or all stock? What percentage? Of gross or net sales? Fixed dollar (or cents) per unit payment? Escalating compensation? Up or down on volume? Options to convert to sellout? At whose discretion?

❏ Other:

You might also address, under certain circumstances, the question of whether or not you would be willing to also accept employment with that firm either temporarily or permanently to carry forward development of your invention. If you have developed tooling and fixtures, would these be conveyed with your deal and if you have developed customer or other intangible relationships, are these included also? At what cost, if any?

FOR INVENTORS WITHIN COMPANIES

Many companies require employees to sign confidentiality and non-disclosure agreements wherein discoveries made by the employee become the property of the employer. Patents granted to the individual are then assigned to the employer, often for "$1.00 and other good and valuable consideration." And this is as it should be; the company after all employs individuals to work for the company.

But inventions of employees which are conceived on the employee's own time and often with his or her own resources are a gray area; sometimes the courts have held for the individual and sometimes for the company. It depends on complex individual circumstances.

Aside from other possibilities for commercializing an invention, there is good reason for an employee to consider his employer as a prospect for an invention. Most companies welcome suggestions from employees and have programs for remuneration. Rewards are available for both substantial and patentable inventions and simple innovative solutions. Find out what your employer's policy is. Perhaps if you have a good new idea, it's marketable to your present employer. You may not have to run off and start your own business after all.

This is a general outline for your proposal but you should also discuss your particular situation with your own attorney or other confidants to be sure you are covering all the features you feel are essential to a satisfactory offer. But be careful! Don't formalize an offer that is so cumbersome and threatening that your prospect will be frightened off. Your proposal should exude benefits and values for the prospect and dispel any natural inclination he might have to view you—an inventor—as a troublesome "crackpot" selling a scheme.

Portray your invention honestly, covering both the benefits and the possible weaknesses. Try to remove every potential element of surprise. Above all, if your invention can't "dazzle them with its brilliance," don't "baffle your prospect with baloney!" It's to nobody's benefit to encourage a new business relationship based on hyperbole or deception. Because "the chickens will come home to roost!"

Obviously, you are not going to prepare such a proposal overnight. If you pursue this course, you'd best prepare for a major effort. When you're finished, you're going to have a handsome proposal, carefully structured and nicely typed (or even typeset) and bound in a presentation folder. And within that folder lies your best chance for selling or licensing your invention. Don't count on a shabby or amateurish proposal for much. Get professional help if you need it!

Now that you have your prospects identified and your presentation in hand, your next step is to approach one or more prospects. There are several ways to start that process.

WORK OUT AN APPROACH

If you're particularly persuasive, glib, and persistent, you might try reaching a decision-maker in one of your prospect companies by phone. This is invariably a tedious process, and if the prospect is out of town, it can become costly. Explain briefly what the purpose of your call is, and try to set up a personal interview. But qualify your prospect first. Determine if they have both manufacturing and marketing capability to fit your product, and if so, would they be interested in another profit center for those resources.

The second method of approach, and perhaps the best, is to write a letter to the prospective company. Using much the same language as I've illustrated above, simply indicate that you are offering for sale (or license) an invention which you believe fits their manufacturing and marketing interests. You may have to experiment with this letter. Your object is to say as little as necessary about the exact nature of the invention, while still gaining enough interest to generate a personal interview.

If you've picked a good prospect, you may get a response, but it's likely to not be what you want! Here comes that "Corporate Moat" again! Many companies have a standard response—a form letter—they send back to inventors and generally it will tell you (1) "we will not consider looking at any outside invention which has not been patented and upon which patent, the inventor relies solely for whatever intellectual property right he or she might have," and (2) "we will not enter into any confidentiality or non-disclosure arrangement as a condition for looking at your invention."

Now I have some very huge problems with that hurdle and if I had a magic wand, I'd certainly make it go away. But if I had that power, I'd make some of life's other menaces go away too. For that matter, is it really that big a problem? True, you are now about to be squeezed down the corridors of a system favoring your prospect, but then you do that every time you fill out a job application or apply for credit, don't you? Life does, after, all have its risks! And perhaps you do have that patent and are confident enough of its strength to not be concerned. And you're confident of having "invented smart" too, aren't you?

TRY FOR A CONFIDENTIALITY AGREEMENT

And if you don't have a patent? How about a filed patent application? Haven't got that done yet either? Then how about your inventor's log book? You have kept one, haven't you? No? For goodness sake, how are any of us to know this is your idea and that you're the author of the work you're trying to license or sell? You know, if you don't practice some "due diligence" about whatever it is you think valuable, how can you expect others to think it has worth? How about this, then: Have you taken the time to make a drawing of your invention and write a description of what it is and does and then have that witnessed or filed with the Document Disclosure Department of the Patent Office? The point here is that if you have been reasonably responsible to your own creation by taking the protective or at least documentation steps available to you, you can have a somewhat more reasonable sense of security in presenting your invention to a prospective buyer, without fear of their abusing your rights.

Some inventors, despite whatever actual patent protection they may have, just won't bargain with any prospective licensee unless the licensee is willing to first sign a confidentiality document and I have seen situations in which a logical business relationship simply could not come to maturity over this issue of confidentiality. Neither party would budge and that was it! Still, it is a matter for you to discuss with your attorney, who may even, as earlier pointed out, act as a "straw" or bridge in the negotiations.

If you license the invention to a company and are to receive a royalty from them, then you will have to arrive at terms. The best royalty terms always provide for a minimum monthly revenue, and preferably are based on a percentage of their gross sales. If the royalty is predicated on net sales or net profit, you have virtually no control over the "burden" of expense they may believe necessary to factor into their overhead. That burden can include a fleet of leased limousines, for example!

This and other details should be worked out with your attorney but a sample licensing agreement is provided in the Appendix of this book.

Keep in mind also that while outright sale or license of your invention may be your preferred option, there is no good reason for not also pursuing other options. For example, consider setting up, even in a small way, to make and sell some of your product yourself. And as illustrated above, the further you carry the product into actual production and marketing, the more definitive it becomes. A limited production and marketing effort converts your invention from a simple idea to a proven product. If that can be done successfully, even on a quite limited scale, your chances of outright sale or license to another company become far greater.

We'll take a good look at what going into production and marketing will involve. First of all, it will probably take money, so we'll explore that in the next chapter.

10.
Financing

The financial "marketplace" is a confusing labyrinth to the inventor looking for funding. You will have to sort through a variety of funding techniques, particularly both collateral and equity funding.

VENTURE CAPITAL

There are many conventional channels for equity funding new businesses. But in recent years, a relatively new kind of company, the Small Business Investment Company or SBIC, has been a real boon to the inventor. It funds people starting new companies, and many of the new companies involve new ideas or products. These companies are partially funded by the Small Business Administration and there are currently about 400 of them around the country. You will find them listed in the Appendix to this book. Note, there is one class of SBIC called a (301) d which serves disadvantaged entrepreneurs such as handicapped persons and members of minority groups.

One of the criticisms leveled against the SBICs is that they are too often interested only in glamor or so called high-tech ventures and many won't look at participations, as they call their involvements, in deals of less than $1 million. Their claim is that it costs them as much to get involved in a $50,000 investment in a new venture as it does to invest $1,000,000 or more. Keep that in mind when approaching a venture capitalist and perhaps I should add "Have no little dreams!"

SEED CAPITAL NETWORKS

Still another source for venture funding is the Seed Capital Network, a relatively new phenomenon in the investment community. These are essentially groups of individuals or companies who have registered their investment interests with a computer databank which provides the service of matching investment opportunities with investors' particular interests. These networks originated, in part, out of studies funded by the SBA which revealed that by far, most new ventures in the United States were actually being funded by private individuals, often known as "Angels."

Most of these Seed Capital Networks operate within limited geographical areas, and since it is a kind of emerging new venture phenomenon, they are sometimes hard to find; you won't easily find them in the Yellow Pages! You will find a list of a dozen or so of these networks in the Appendix of this book, however.

Almost all of both kinds of venture capital source specialize in risk ventures and, since risk is the category most new inventions belong in, it's a welcome opportunity for the aspiring inventor.

BUSINESS PLAN The venture capitalist will want from you a detailed business and marketing plan together with a personal resume. And he'll want to see "living proof" of your invention, not drawings or clumsy wooden dummies, but honest-to-goodness working objects, preferably in the form in which you expect to bring them to the marketplace.

Then, your venture capitalist will want to examine your costs and pricing. You can be sure he will be knowledgeable about what "formulas" ought to apply. He'll want the estimates figured out to the last detail. In the accounting, he'll want to see a "cash flow" projection, your expert predictions of what kind of funding will be needed. He'll want to know about machinery, workplace, employees needed, marketing strategy, advertising budgets, freight factors, and, the competition!

In short, you're going to need to prepare a complete plan, fully informative and as honestly objective as possible. This plan may be crucial to your success. Remember, venture capitalists are looking for opportunity, but they don't buy pipe dreams or wishful thinking. If anything, your plan should err on the side of conservatism when you make broad assumptions about market possibilities.

The following information is extracted in part from resources supplied by the Small Business Administration as is the compilation of firms offering investment services.

OBTAINING SBIC FINANCING FOR YOUR SMALL BUSINESS SBICs exist to supply equity capital, long-term loans, and management assistance to qualifying small businesses.

What Types of Businesses Qualify? They invest in all types of manufacturing and service industries. Many investment companies seek out small businesses with new products or services, because of the strong growth potential of such firms. Some SBICs specialize in the field in which their management has special knowledge or competency. Most, however, consider a wide variety of investment opportunities.

Only firms defined by the SBA as 'small' are eligible for SBIC financing. The SBA defines small businesses as companies whose net worth is $6.0 million or less, and whose average net (after tax) income for the preceding two years does not exceed $2 million. For business in those industries for which the above standards are too low, alternative size standards are available. In determining whether a business qualifies, its parents, subsidiaries, and affiliates must also be considered.

Approaching an SBIC   If you own or operate a small business and would like to obtain SBIC financing, you should first identify and investigate existing SBICs which may be interested in financing your company. Try to learn as much as possible about those SBICs in your area, or in other areas important to your company's needs. In choosing an SBIC, consider the types of investments it makes, how much money is available for investment, and how much will be available in the future. You should also consider whether or not the SBIC can offer you management services appropriate to your needs . . .

Plan in Advance   You should research SBICs to determine your company's needs well in advance—long before you will actually need the money. Your research will take time.

PREPARING A PROSPECTUS/
BUSINESS PLAN   When you've identified the SBICs you think are best suited for financing for your company, you'll need to prepare for a presentation. Your initial presentation will play a major role in your success in obtaining financing. It is up to you to demonstrate that an investment in your firm is worthwhile. The best way to achieve this is to present a detailed and comprehensive business plan or prospectus. You should include at the minimum the following information about your business:

Identification. The name of the business as it appears on the official records of the state or community in which it operates.

The city, county and state of the principal location and any branch offices or facilities.

The form of business organization and, if a corporation, the date and state of incorporation.

Product or Service. A description of the business performed, including the principal products sold or services rendered.

A history of the general development of the products and/or services during the past five years (or since inception).

Information about the relative importance of each principal product or service to the volume of the business and to its profits.

Marketing. Detailed information about your business's customer-base, including potential customers. Indicate the percentage of gross revenue accounted for by your five largest customers.

A marketing survey and/or economic feasibility study.

A description of the distribution system by which the products or services are provided to customers.

Competition. A descriptive summary of the competitive conditions in the industry in which your business is engaged, including your concern's position relative to its largest and smallest competitors.

A full explanation and summary of your business's pricing policies.

Management. Brief resumes of the business's management personnel and principal owners, including their ages, education, and business experience.

Banking, business, and personal references for each member of management and for the principal owners.

Financial Statements. Balance sheets and profit and loss statements for the last three fiscal years or from your business's inception.

Detailed projections of revenues, expenses, and net earnings for the coming year.

A statement of the amount of funding you are requesting and the time requirement for the funds.

The reasons for your request for funds and a description of the proposed uses.

A description of the benefits you expect your business to gain from the financing—improvement in financial position, increases in revenues, expense reduction, increase in efficiency.

Product Facilities and Property. Description of real and physical property and adaptability to new or existing business venture.

Description of technical attributes of production facilities.

In other words, you'll have to do a thorough job preparing this proposal. But then this is why we covered a lot of those raw necessities in previous chapters. In addition, study the sample New Business Proposal in the Appendix. Note also the accompanying financial statements. And remember too, that writing your proposal is important not only when you seek funding, but also to outline your own plan in a businesslike manner. And again, the same kinds of information will be important in a proposal to sell or license the invention rather than seek funding for a manufacturing and marketing venture.

WORST POSSIBLE SCENARIO

A venture capitalist wants to know about the "worst possible" scenario. Thus, for both him and yourself, you must structure your plan accordingly. Here again, your original cost accounting will prove invaluable. In fact, having worked through all the tedium of the first few chapters of this book, you should have most of your information available for writing your business plan.

Finding a venture capitalist is something like hunting for a job. You dress up for it, get a good night's rest, shine your shoes and go knocking on doors. In large measure, YOU are going to be what they might invest in. What is your credit rating? What business experience have you had? Do you make a good sales presentation? Even your personal habits will come under scrutiny. If it's a non-smoking office, better not light up! If you go to lunch, skip the drinks! Know when to be serious, and when to "slacken up."

My experience with venture capitalists has taught me that in general, they are extremely capable. Most are well trained in financial and legal disciplines. As long as you present yourself as a reliable and professional "custodian" of the invention, you'll be treated fairly.

If the "numbers add up" and the venture capitalist likes the proposal, he'll suggest a follow-up meeting. If your proposal doesn't fit the kind of "portfolio" of investments they are looking for, he'll say so. And you'll have to knock on some other doors.

If you do get that second interview, it may be at your office or plant or garage, wherever your endeavor is located. This meeting may be simply an extension of your first interview, to give him further insight into you and your project. This time, he may offer you a proposal.

START-UP PHASE

The proposal will usually sector the venture into phases: a "start-up" phase and a public funding rollover phase. For example, if you need $100,000 to launch your company, they may offer to put up that entire amount for an initial stock ownership of 20 percent. The contribution you make for your 80 percent will be your invention, your talent, and hard work. The cash input may be incremental, deposited into your checking account in amounts just sufficient to cover your equipment acquisition, inventory of raw materials, and other expenses. Do not worry about that. It's part of their cash management to do things this way. Good investors don't leave money lying idle in non-interest bearing checking accounts! On the other hand, it wouldn't be out of order for you to ask for the whole lump sum—and let your new enterprise earn the interest.

As part of the deal, they will help you with the incorporation of your company and the details of setting up the business package. They will place one or more people of their own choosing on your board of directors. Often, the venture capitalist will provide someone to become treasurer and also bookkeeper for your company. This involvement gives them inside information on your conduct of business. Equally as important, their input can give you valuable guidance in the day-to-day running of your new "empire."

MAINTAINING THE PLAN

The venture capitalist will want you to maintain the business plan you outlined in your initial presentation. Their investment is in that particular plan and they expect you to follow through with it. If eventually you discover something "out there" you had not foreseen, don't just adapt to it. Go to your investor first and present the problem to him. If it means a change in your plan of operations, the venture capitalist has the right to participate in the decision-making process.

However, sometimes there are valid reasons for switching courses, not just to survive but to actually profit from unanticipated discoveries. The giant 3M company made dramatic shifts in its original business plan because management could not have foreseen that a gooey substance in a test tube would become the basis for "Scotch Tape." Today, 3M has about 50,000 products in its line! Not bad for a company with plans for a sand pit. Most likely, no shift you make in your business plan will be that dramatic or that profitable. But if you think you have a new idea that could help your company's future, you should present the idea to your investor.

PUBLIC STOCK OFFERING

At a certain point, when you've achieved success, or at least enough to know that your idea is standing up to the scrutiny of the marketplace, the second phase of funding may become necessary—that of a public stock offering. The venture capitalist knows about this kind of funding and will be able to take you over this next big hurdle, if that course recommends itself. It's called an IPO, or Initial Public Offering and it means selling your securities to the general public. For this you must have Securities and Exchange Commission (SEC) approval.

Remember that two extremely important factors influence a venture capitalist's interest in a start-up: management and profitability. He'll want to know about you, and any team of management people you may have enlisted for the project. As a rule of thumb, the venture capitalist will want to see a carefully constructed profitability projection demonstrating that the net value of the company can increase ten times in five years.

In general, the venture capitalist is a likely and helpful source for funding a new product or technology-based company. To some extent, he is prepared to see you through the troublesome waters of the start-up. For that, he's entitled to a reasonable chunk of your company. However, you may want to consider other forms of funding as well.

Before the rise of venture capitalist companies, the equity (investment or risk-taking) funding of companies was usually handled by large investment banking companies and stock brokerage firms. A company of this size and stature might be interested in funding your start-up. But, usually, these firms prefer to deal with seasoned people. Moreover, your invention may not require the amount of capital these people prefer to deal in.

MERCHANT BANKING

Lately, many conventional banks have begun experimenting with what is known as merchant banking. In place of the traditional asset-based financing, they have become involved in "participation" financing. Remember that banks are under considerable restraint by the government to be judicious custodians of other people's money. If you're a savings account holder, you can appreciate that. But the bank must make a profit on its operations to pay interest on these savings. Many of them, looking at their investment portfolios find a heavy percentage "invested" or loaned out on aging packages which bear a nominal return in today's market.

DEVELOPMENT COMPANIES

Given all of that, the bank is looking for high yield opportunities. Thus, bankers in many communities are seeking ways to help build their "industrial base." This means jobs, more business for local merchants, and taxes. These banks often become tangentially involved, then, in industrial development companies, and you should discuss his with either your local banker or the Chamber of Commerce. You may have exactly what they're looking for. And they may have a lot to offer you.

INSURANCE COMPANIES

One source for funding often overlooked is the insurance company. Like bankers, these firms are also looking for investments. While most insurance companies work through the large investment houses in placing funds, you may at least investigate whether one of the insurance underwriters could be interested. However, like the larger investment houses mentioned earlier, insurance companies are not set up to "baby sit" new ventures. Unless your idea has the magnitude of an anti-gravity machine, don't get up your hopes!

Initially, you may have to rely on your own resources or those of friends and family members. Of course, others' investments rise or fall with the success of your company. If they are willing to risk their capital on your invention, they have certainly earned a share of any profits you make.

Once you claw your way into a production stance, sell and deliver some goods and have an accounts receivable of well-rated clients on the books, you can factor the accounts receivable. Usually this means assigning the receipts to the lending institution, the factor. You get money up-front (to produce and sell more goods), and, when the payments come in, you pay back the creditor.

Often, the receivables are assigned to the factor "with recourse," meaning that if the account doesn't pay, you make it up. Generally, depending on your strength and the quality of your accounts, you receive only 75 or 80 percent of the billing in cash up front. A reserve of 20 or 25 percent is held back as a kind of insurance for the factor. Interest is high and it is an expensive way to do business. Preferably, you will want to fund your effort more conventionally right at the start-up stage of your company.

Your local banker will get into the picture on a similar basis once you have some kind of track record. Moreover, your banker, particularly if he is a commercial banker, will consider financing equipment, raw materials and other conventional requirements. Early on, establish a good relationship with your banker, regardless of what other start-up capital you raise.

Financial difficulty is, by far, the most trying experience of conducting a business, and can demoralize and virtually disable a young company. If your planning was right in the beginning—if you were not overly optimistic in your sales predictions, if you were realistic about your costs and time scheduling—you might not run out of money. But more than a few do. If it happens to you, get busy immediately to remedy the problem. If you don't remedy it, you'll soon be out of business.

Probably the very best course, in a cash-short situation, is to "face the music" with your creditors. Don't put it off. Don't wait for wheels to begin squeaking. Go to your creditors, explain your problem and your plan for correcting it! Discuss the situation up-front. Be honest. But be realistic too. If necessary, offer them additional security, a personal note, or collateral such as finished inventory.

FINANCIAL PROBLEMS

For many new businesspersons, facing tough financial problems is a frightening experience. But I can't emphasize too strongly how important it is to communicate. The business community is usually willing to work with struggling companies if their managers are honest about those financial struggles. But there is little tolerance for entrepreneurs who won't answer mail or telephone calls. Such cowardly management is less tolerable than management with a weak capital position.

In my early experiences with bankers and investment capitalists, I made some costly mistakes. I harbored, for one thing, an unhealthy attitude toward them. Money is arrogant, I thought. Later on I discovered that "arrogant" was an inappropriate term. Pragmatic, yes, but not arrogant. What venture capitalists and others are tuned to is the bottom line—profit! They are not philanthropists. That's another business entirely.

Another mistake I made was to shower the funding company with interminable reports about everything I was doing or expected to do. They didn't want all that detail, just a healthy balance sheet. I learned that reinventing the wheel was to be avoided. Stick with the original plan and keep your reports simple, your objective clear!

Most new companies fail for lack of capital and management. If you have a good idea and management capability too, you can find the capital. If you can't convince an investor that both the idea and your capability are sound, you might have to shelve your ambitions until you have the "right stuff." Perhaps you must take a little more time to do it alone. But don't give up too quickly. Many of the best commercial successes were born of many battles. Persistence is the name of the game!

11.
Manufacturing

Recently, I prototyped a product for an inventor who was also quite successful in another line of work. For months, we had been flirting with the inevitable step of going into production. We had reached the point of manufacturing tooling. We went over the figures again and again, considering all the worst possible scenarios. But somehow, he just couldn't make that important leap.

PRODUCTION REALITIES

Finally, he flushed slightly, squirmed a little and asked, "What is this leading me into? What kind of money am I really talking about in going into this business?" The truth was out! And he had to face it! Was he really prepared for the realities of setting up to manufacture his product? Did he have the time for it? How would it affect his lifestyle? Did it mean giving up his other profitable employment?

These are problems an inventor must face at the outset and they have nothing to do with the purely mechanical requirements of a manufacturing operation. They have to do with whether the enterprise can be operated as a comfortable little endeavor in the garage or as a full scale factory layout.

SALES-DRIVEN PRINCIPLE

Perhaps it is my own stringent (even penurious) attitude, but I am a great believer in operating a company on the sales driven principle. In my own experience, it is easier to handle start-up, alibiing about late delivery, than it is to make installment payments on equipment gathering dust! That is exactly why the chapter on promotion preceded this one!

So at this point, I am going to assume you have generated some sales interest if not outright orders. I hope you have hedged the delivery dates on the orders to accommodate the inevitable. But hopefully, you have also, through your prototyping stage, developed some inkling of what it will take to actualize your manufacturing step. And there may already be some equipment at hand with which to start.

CONTRACT MANUFACTURING

Because there is so much to know about any given manufacturing process, I strongly urge the inventor to rely on contract manufacturing as much as possible at the start-up stage. At least, if funds are limited. Once you have sufficient funds, you can do many things differently, including employment of skilled specialists.

You can learn much by dealing with the contract manufacturer. As time goes on, you can become knowledgeable about the standards mentioned above, and about the specific problems related to your product. You may pay a little more at the start, but you do not have the unnecessary burden of your own manufacturing plant. As you develop knowledge and experience, you may then take over one or more elements of the manufacturing and lower your costs.

Various elements in the manufacturing process become more clearly defined, and when you take over those particular steps, you are able to isolate them from the more costly and less predictable start-up phase of manufacturing.

Many firms provide this service, either as a totally independent activity, or in connection with their own product line. Back to the Yellow Pages! If your product is mostly plastic, go to a plastic company specializing in one or more of the techniques required for your product. Avoid separating the tooling requirements from the production. Get single-source responsibility for a particular part.

SINGLE-SOURCE QUOTES

The same thing applies to metal parts. Get a single-source quote and responsibility for the patterns and part production. That is, unless you have experts in all the areas of manufacturing you use—pattern-making, the foundry, etc.—you should not self-contract with several sources. The parts manufactured by one source may be incompatible with the manufacturing expectations of another source.

If you need machined parts, see if they can be produced by someone specializing in that kind of work. For example, some companies do nothing but produce "Swiss screw machine" parts—things like small ferrules, screws and landed shapes. Other firms do nothing but "centerless grinding," "EDM" machining, or molding tool "texturing." You should, if this specialization fits the requirements of your invention, familiarize yourself with as many of these techniques as possible.

In other fields, such as chemical processing, you will find firms who also do contract manufacturing. There are firms which specialize in mixing and blending ingredients, packaging, tableting, and many other things. Many of these firms also offer the added advantage of a "quality control" (QC) laboratory.

FIRST ARTICLE APPROVAL

When ordering contract manufacture, it is customary to specify "first article approval." This means that pilot manufacturing is set up and production-quality parts are run. It is your responsibility to measure, test and approve the part. Most contract manufacturing firms have a "setup" charge for this, but do not avoid the responsibility. I've supervised start-up production in which the client's engineer stood by to "QC" parts as they began coming off the line. If the parts were "to specification," he "signed off" which meant giving the go ahead. The manufacturer retained an approved sample, so that continuing production could be cross-checked for quality consistency.

STANDARDS

For virtually all manufacturing processes, there are STANDARDS. For example, in the printing industry there are standards for color and these are known as "PMS" standards. When you talk about red, you can specify exactly what red you want, by referring to a specific color swatch in the PMS color sample book. The use of standards eliminates a lot of misunderstanding between buyer and producer.

In machining, there are "Rockwell" and "Brinnel" standards for material hardness, "Durometer" for flexible materials, "SAE" and "ISO" standards for machine threads, government "MIL Specs" for nearly everything, "RMS" standards for surface finish, standard screen meshes for particle sizing, tensile, modulus of elasticity and flexural tests for other materials.

These are but a very few in just one narrow range of the manufacturing discipline. New standards are developed constantly as new materials are introduced. For this reason, the independent inventor should rely on a reputable contract manufacturing source for as much of the initial start-up as possible.

FUNCTIONAL INTERFACING

Many of the problems of start-up have to do with poorly defining the requirements for a part or assembly. If it is what is called a "free standing" part, that is, complete as a saleable object, the tolerances or standards may not be as stringent. But when several parts go into the final device, you are faced with what is jokingly called the "Canaveral" law of physics: "If it can go wrong, it will." In functional interfacing, the magnitude of potential misfit is startling. If two pieces must fit precisely, the possibility of misfit is 50/50. But if you must join ten parts, for example, a train of gears, the possibility for error is on the magnitude of ten to the tenth power. This is also called a "fault tree" in design engineering.

THINK ABOUT IT
If you can do it once, you can do it a million times. *But is this what you really want to do?*

Obviously, in defining the manufacturing process, it's important to keep these odds in mind. The finished device may simply not work! In the prototyping stage of product development, these realities are often overlooked. Their omission translates into disaster in the manufacturing step.

CONTRACT ASSEMBLY   In many communities throughout the country, there are handicap workshops which serve the dual purpose of offering contract assembly and packaging services while at the same time giving meaningful employment to the disabled. Many firms employ primarily elderly persons whose sense of craftsmanship and performance is excellent. When putting together your manufacturing plan, definitely explore this option.

If you decide to do your own manufacturing, explore the used equipment dealers first, and be inventive about the machines you buy to fit your needs. I have found unlikely sources for a piece of gear. Once, I purchased an old butter churn to use as a mixing machine for a blended product. It worked extremely well for the purpose.

EQUIPMENT   If possible, avoid single-purpose equipment. Don't build a special conveyor if you can buy a conventional one. Don't design a special tank if you can buy an existing model. You may have to modify it some, but even in doing that, try to make the modification reversible. For one thing, it may not work so you may want to revise your modifications. Second, you can later remove the modification and sell the standard tank on the used equipment market.

Many new entrepreneurs get carried away with "brick and mortar." They assume a fancy office and a gleaming plant will improve their image to such an extent that business will flood in. Later they realize it doesn't work. My philosophy is that if your invention is successful and the profits roll in, let the profit pay for the little extra niceties but only after you've trimmed the sails in the manufacturing setup.

Whether your manufacturing begins on your kitchen table or in a bona fide plant, there is nothing like the thrill of seeing your "baby" come off the production line. At this point, you have the opportunity to compare your actual costs and profits with your front-end calculations. This is no longer conjecture, it's reality.

Now you can put some more of your inventive talent to work, smoothing out the little bugs, speeding up your process, eliminating unnecessary steps and materials. Sometimes, as in chemical manufacturing, you can "dovetail" processes to improve efficiency or explore by-product recovery.

In one such manufacturing "tune-up" I took waste heat from a cooking step, increased the drying efficiency in another machine and eliminated vapor lock in the boiler feed water, all with a 50-foot coil of copper tubing and an old galvanized tank. It was a thrill equivalent to the invention we were processing!

DEVELOPING A SETUP In developing a manufacturing setup, keep these things in mind. Make the setup flexible, to accommodate improved efficiency, product redesign or model changes. Design the manufacturing setup so that it may be up-scaled, that is, translated identically into a larger version of the same thing, or an exact multiplication of the original. Sometimes, manufacturing processes can be up-scaled with greater efficiency or better quality output and sometimes not.

Carefully examine the cost components of your manufacturing steps. Some steps may be "labor intensive" and can be automated for greater efficiency. Often the automation is nothing more than some appropriate jigging that will speed and improve the accuracy of the work. I'd be aware of the "robotic" revolution but careful about acquiring an "android" for a manufacturing step which may prove obsolete down the line.

BE FLEXIBLE Be flexible in your judgment about how a thing should be done. Let the worker figure it out, or do it yourself and find out the easiest way to finish the task. In time, you will discover an efficiency of hand and arm movement, best positioning of a piece, where the manufacturing bottlenecks are, and the like. Then you can define the steps formally, and consider how mechanization will improve the process.

STANDARDS ACT Be extremely alert to potential manufacturing hazards. Even a simple machine can quickly amputate an unsuspecting worker's hand. Or worse! The Occupational Safety Standards Act requires a safe working environment. This means enclosures for moving machinery parts, railing guards, machinery safety interlocks, slip proof floors, and much more. The penalties are severe for any manufacturer ignoring the rules. But the benefits in decreased industrial accidents are worth it. Moreover, a safe workplace will add up to improved morale and greater efficiency. And that means profit!

LIABILITY INSURANCE This is perhaps the best place to comment about the matter of insurance. Today, manufacturers are held more liable then ever before in the marketplace. You must assume that if it is even remotely possible for your product to cause injury of any kind to either the purchaser or anyone else, you may be held liable. Even if you plaster your product with disclaimers and warning labels, you will still be held responsible!

So find a good insurance man, representing a reputable underwriter experienced in product liability. He will examine your product carefully and then quote a price. Don't be afraid to ask questions, and to

---

WHAT IS QUALITY?

Too many American manufacturers miss the point entirely. Consider these examples:

A major electric tool company is told by the consulting psychologist that the roar of a power tool infers brute force and power. The company proceeds, therefore, to produce an electric sander that literally screams.

A principal food company, in its zeal to produce a new low-cholesterol butter substitute, has just introduced a table spread which one consumer likened to "axle grease."

A venerable small appliance company recently introduced a low-priced toaster in which the heating elements are so drastically under specified that bread is dried out before it toasts.

As inventors, we should never participate in such foolishness; it is a shoddy sacrifice of American quality on the altar of dimwittedness!

---

take his advice. Many companies who specialize in this type of coverage are highly involved in maintaining product safety. They may recognize product hazards that you had not even considered. You can improve the safety of your product before it goes on the market, reduce your liability risk and lower your insurance cost. More importantly, you may have avoided a possible tragedy!

12.
Find a Market and Serve It.

By far, the most formidable challenge for the inventor is marketing. The old adage about the "world beating a path to your doorstep" is simply not true. Products do not sell unless people know about them. Too often, the inventor is gripped with the notion that if only he could get a few samples of his product, he'd be off and running. Unfortunately, it just doesn't work that way.

To market a product one must determine what the product benefits are, who the audience is, why they should buy the product, and where, when, how, and at what price. The marketing man's language is that of "demographics," "market share," and "psychological inducement."

What is a market, anyway? There is a lot of disagreement about even the term. It can be a neighborhood, or a continent. It can be a particular age or ethnic group, wholesalers, retailers or consumers. It can be "vertical" or "horizontal." It can be beer drinkers or crochet lovers, anglers or garden clubs. It can be a few, or millions. And it all depends on what you're trying to sell.

MARKETING SERVICES As a lone inventor, you have several choices to make when assessing your market. To help you recognize those choices, you may want to retain a consulting firm, usually listed in the Yellow Pages under Business Consultants or Marketing Consultants. Expect to pay a substantial fee, but also expect to receive some highly professional guidance.

But a word of advice concerning marketing firms: Beware of firms which offer "Invention Development and Marketing" services. While some such companies may offer useful service, many are the extensions of manufacturing firms which are simply trying to further the profitability of existing tool rooms or marketing organizations. Often, their expertise is limited. If the parent company is in the hand tool business, they may be able to assist with certain aspects of your manufacturing, but have no expertise in marketing your product if it is not a hand tool.

Before enlisting the services of any invention development or marketing firm, be sure to check with the local Better Business Bureau. Most bureau offices can provide information on the reliability of specific companies. Some states have specific laws protecting the unwary inventor against unscrupulous operators. For example, the state of Minnesota requires firms to provide the consumer (inventor) with the following information:

❑ A full statement of the contract price, a three day cancellation period, and a cautionary warning concerning legal advice on patent matters.

❑ A full description of the service to be rendered.

❑ A completion date for when the service will be rendered.

❑ A statement specifying whether the developer will acquire a legal interest in the inventor's invention.

❑ Statistics showing the number of customers who have profited from the developer's service.

❑ Invention developers must post a bond with the Secretary of State.

A truly professional marketing firm will examine your product carefully and suggest a plan for presenting the item to the public. The plan will include an appraisal of any necessary consumer interest, distribution channels to reach those consumers, packaging and pricing. A good market research team will examine, among other things, the competitive "influences" in the marketplace and possible advertising programs.

Such probing helps to define things like "perceived value," factors which would motivate the sale, where the consumer expects to find the article in the stores, and what the consumer expects for his money. These assessments, while sometimes brutally revealing, are important, if the product is to survive the rigors of the marketplace. Many marketing firms retain "focus groups" of ordinary, walk of life people, and your product is submitted to these people for feedback and criticism. For example, toy manufacturers often employ child focus groups to test not only consumer interest, but safety features. Many toys, conceived by adults, do not pass the child test!

PACKAGING IS CRITICAL

For many products, packaging is a crucial part of the product. Once, packaging was essentially a containing, protecting and shipping feature. No more. Today the packaging must also sell, because most stores have become self-service "warehouses." The package must have good point-of-purchase appeal. In addition, it should be engineered to discourage pilferage. At one time, lipsticks were displayed "raw" in open store cases. Today, they are blister packaged on cards which are larger than an average pocket!

Some products sell best within "companion specie" settings, while others are best in "impulse" settings, such as at the cash register. Certain colors, appropriate for one kind of product, destroy sales of another. This is where the marketing consultant can really help.

VARIED DISTRIBUTION CHANNELS

Today's marketplace is not the simple thing it once was. Before chain stores, there were essentially wholesalers and retailers. In between, there were usually brokers or manufacturer's reps who represented manufacturing clients and mostly called on wholesalers. The pricing and markup percentages were relatively well defined, as were brokerage commissions. It was a simpler life for the small start-up company.

For certain kinds of merchandise, a third kind of firm, called the jobber, wagon jobber, or stocking distributor actually purchases merchandise, like the distributor or wholesaler; but the jobber specializes in placing merchandise directly into stores, right on the spot. Typical are firms carrying goods like tobacco, candy and snack items. Also, there are still many jobbers who specialize in tools only.

Today, the marketplace is not so clear cut. Large groups of retail stores have banded together into buying co-ops. Chains buy in large volume and insist on wholesale pricing or better. Department stores, with multiple locations, do likewise. Brokers and manufacturer's reps still, however, represent manufacturers to this redefined marketplace. There is an even greater need for local, on-the-scene representation. Buyers for merchandise in department stores are no longer just buyers, but are merchandise managers, held accountable for not just the wholesale purchase of goods, but its resale to the consumer.

In modern merchandising, very little buying is done by individuals. Instead, it is done by buying committees. Retail selling has become a well-defined science. The bottom line is "How much profit can be generated per cubic foot of display space?" Everything is on a computer which constantly monitors inventory, calculates "spin" (the frequency of stock turnover), profit yield and the like. When you consider that even an average sized variety type store will carry more than 100,000 different items, exclusive of size and color, it is easier to understand the enormity of the retail store's task and thus, the buyers' discernment when adding new products to their shelves.

One company, Wal Mart stores of Bentonville, Arkansas, has a well-publicized policy of buying from American suppliers. When you walk into a Wal Mart store you will see a sign at the entrance inviting you to submit offers on any product you see on their shelves—made off-shore—which you believe you can manufacture and price competitively. This is, in my estimation, outstanding—and good news for American inventor/entrepreneurs.

Other large companies are reluctant to discuss their buying policies. Target Stores, Inc., of Minneapolis, Minnesota declined an interview with me and instead submitted a document entitled "Criteria for Vendor Evaluation - Target Vendor Recognition." Here is the content of that document:

Dependability

- ❑ Ships on time and complete
- ❑ Anticipates and communicates problems up front
- ❑ Follows through on information, samples, paperwork
- ❑ Meets deadlines
- ❑ Renders credible information

Cooperation

- ❑ Partnership with growth being the goal
- ❑ Recognizes the importance of regional needs to Target
- ❑ Flexible on shipping dates to meet distribution needs
- ❑ Shops our competition—feeds us information to help us grow
- ❑ Good rapport—easy to work with sales staff

Shared Risk

- ❑ Backs up stock in reserve against our ads
- ❑ Financial commitment (return policy, markdown allowance, manufacturers' rebates)

Industry Leadership

- ❑ A leader not a follower
- ❑ A vendor other vendors watch
- ❑ Constant flow of new product development and packaging
- ❑ Travels and evaluates marketplace for product and presentation development

Creativity

- ❑ Supports new product development
- ❑ Experiments with new treatments in color and design
- ❑ Develops a marketing plan to compliment and exploit a new item

❑ Able to adapt a product to meet our customer's need and our distribution needs

Financial Contribution

❑ Sales growth

❑ Profit growth

❑ Participates and gives concessions in the areas of discounts, advertising and dating

❑ Covers testing and planogramming charges

Quality

❑ Shares the pride in selling a quality product

❑ Improves the quality both at our request and on their own

❑ Takes care in packing and shipping

❑ Respect of and intent to put the quality in the product

Note: It is not mandatory that a vendor fulfill every requirement, but rather the majority.

Difficult as it seems to obtain good communication with some retail firms, it is important nonetheless to recognize that they cannot survive without good quality products to sell, and a steady flow of new and improved ones too. Product improvements—which inventors and innovators provide—are after all by far the most important sales leverages available. How much better to be able to offer an improved product to increase sales, than hold a sale and discount merchandise to move it off the shelf.

Still there are general rules for approaching large companies. Here, from the Inventor's Resource Directory, is a passage from a section on retail marketplaces:

COMMUNICATING WITH BUYERS

1. Address correspondence to "Buyer: (type of merchandise)" followed by the company address. Most companies prefer this, until such time as a business relationship evolves and negotiations are conducted then with a specific buyer by name.

2. Unless you are seeking an appointment for a personal sales call, do not telephone companies to engage them in idle inquiry.

3. If you solicit business by mail, your communication must include at least the following:

A. Name and brief description of the product.

B. Catalog sheet illustrating use and purpose of the product, and a brief description of the packaging and master case pack.

C. Current price sheet giving quantity and trade or payment discounts.

D. Description of manner of handling freight charges. (Does the product price include freight or postage? Do you prepay freight charges and bill it back with your invoice? Does the customer pay freight charges collect?)

E. Will you "drop ship" to individual stores? Is there an extra handling charge for this?

F. Advertising or promotional incentives being offered.

G. State and proof of public liability insurance indemnification.

H. The full name, address, phone number and contact person in your company.

I. Financial references and whatever other brief materials you may submit to demonstrate your responsibility to quality, delivery and service.

4. Do not submit samples unless requested to do so.

THE MAIL ORDER ROUTE

A fast growing segment of retailing is in mail order, either directly through display or classified advertising in the appropriate magazines, or through one of the reputable mail order firms who publish catalogs. This may be the marketplace you'll want. It is easy to get into with just a classified ad and there are national publications serving virtually every special interest. Check the business section of your library and study *Ayer's Guide* for a comprehensive listing of all major publications. For most listings, circulation is given, and sometimes a profile of their subscriber audience. This information will give you insight into the potential size and quality of your perceived marketplace.

The marketing consultant will check mailing list directories and try to pinpoint the number of wholesaler, jobber, retail or other marketplace buyers who would be candidates for your product. Virtually every conceivable market has an SIC (Standard Industry Classification).

The Federal Department of Commerce and local chamber of commerce can often provide useful insight into candidate markets. Even a cursory check of the Yellow Pages of a large metropolitan community may prove helpful, not only in terms of your buying prospects, but also of competition you may not be aware of.

MARKET TESTING   All major companies market test. So should you! In a sense, it is a kind of "prototyping" exercise, only you will be testing market feasibility rather than the functionality of your invention. Market testing should test out a specific marketing strategy which can then be "upscaled" to a bigger effort later on. It is a cautious way to tackle the unknowable.

Let's suppose you have a new kind of can opener. You believe, from your earlier market assessment, that it belongs in the housewares department of a department store. The first thing to do is go to the manager of that store, and explain what you are trying to do. In this case, you want to find out if your public presentation will entice consumer buying. You're not trying to dump a carload on him or her, you just want to do a market test.

At this point, you may run into your first objection. The manager may tell you he doesn't like the packaging, for example. If your marketing consultant did his job in the earlier stages, this shouldn't come up, because that is part of assessing a market.

If you get a cooperative ear, the manager will look around his department and perhaps suggest how he would display it. You may be in for a shock! Working as you have, through all this invention custodianship, when you see your package displayed for the first time, you may think, "How puny it looks among all that other flashy stuff!" It well may. One of the pitfalls is that in developing a new product, it grows larger than life. You see it in a different perspective when it's placed into its point-of-purchase environment.

This is one of the reasons for assessing a marketplace. There is no substitute for going directly to that environment and studying not only similar products, but all the products surrounding them. The producer of every item wants his product to attract the consumer's attention. And you are competing in a large marketplace.

You may have to "consign" the product to the store to get it displayed. You may even have to give it to the store! Be prepared for that. All you want, at this point, is some public indication of salability. At any rate, get some product into the store, or some store, and then see what happens.

> ### WHAT TO LOOK FOR IN A SALES AGENT
>
> If you elect to pursue independent commission sales representatives to market your new product, here are some considerations:
>
> 1. What exact geographical territory does the candidate firm cover?
> 2. With what frequency do they cover the entire territory?
> 3. How many sales employees does the representative have making calls?
> 4. Do they carry competitive lines?
> 5. What related lines do they carry?
> 6. Names of other principals represented?
> 7. Can they be contacted for references?
> 8. What companies and buyers does the representative deal with?
> 9. Can they be contacted for references?
> 10. How long have they been in business?
> 11. Does the representative offer ancillary service such as store detailing, product demonstrations or technical support?
> 12. Does the representative offer warehousing services?
> 13. Does the representative offer shipping and billing services?

At this point, since you are market-testing the bare product, do not attempt any special point-of-purchase advertising. In fact, you should request conventional shelf location, rather than an end-aisle or high promotion location. That comes later. For now, you want to test the point-of-purchase appeal of your item.

This is called "rough tuning." If this step goes well, if people buy your product, then you will want to test your promotion ideas. Get feedback from the store manager. Is he satisfied with the movement? Are there any complaints? What does he suggest for additional promotion? Does it also need newspaper or other media advertising? Store banners? "Two for one" deals? Rebates? Listen to him. He's your link to the consumer!

In getting this kind of market test, you may have to "go to the top." Most large stores do not permit department managers to deal directly with salespersons. They'd be besieged. You will probably have to go to the buying office of the retail organization and present your plan to them. Many large retailers have test stores for assessing new products. The company headquarters may well have a demographic portrait of the clientele of the test store, that is, whether their customers are housewives, students, businesspeople, and so on. When they test a supplier's product, they can develop projections to determine how well the new item will do in other stores.

Retailers are always looking for new items, even though getting the new item into their system is a big task. They look for new items because almost nothing sells a product with greater forcefulness than the word NEW. So your new product has that going for it at least!

GETTING MARKET FEEDBACK

But your product may not be a consumer item in the above sense. For example, you may have a new device for improving the function of a machine in an industrial plant. Then, you must go to a plant with that kind of machine, talk to the purchasing agent and ask for an opportunity to install your device on their equipment to see if it actually does speed production and improve quality. You want the people working on the machine and the quality-control room to assess your device.

Listen carefully to their advice. If they don't like your device, ask why. Ask them what they would like to see improved. But be careful. Sometimes you can be set on a course of reinventing your product needlessly, just to suit the whim of a single feedback. When you get negative feedback, test the validity of the opinion elsewhere. If you get the same suggestion twice or more, you will want to consider it.

You may also want to test the market through mail order. Regardless of whether you choose a general publication (a newspaper or general interest magazine), or a special market publication (a trade journal), you'll be testing for several things: the item's appeal, the appeal of your advertisement, the acceptance by a specific group of buyers, and the efficacy of the publication you chose. If any one of these flop, the whole thing flops. And you don't always know which one or more factors were the cause of the failure. Likewise, you may not gain insight into why it succeeds. So mail order, while a relatively inexpensive way to target a buying audience, is not necessarily a good way to test the waters.

There is a saying in the profession that "Half of all advertising is no good. Trouble is we don't know which half!" The purpose of market testing is to eliminate as much of the wasteful half as possible. It's called "targeting" your audience. The objective is to eliminate as many of these inefficiencies as possible, literally before you go broke!

During your market testing stage, your product will go through a lot of growth. You may have to make design and packaging changes. But if you've done your homework early on, you won't have to give it up. Just hone it down and smooth it out.

If possible, test your consumer product in several regional marketplaces. Many products find primary acceptance in marketplaces far away from the inventor's home town. And even though you expect your item to sell best in a department store, you should test other places, such as a hardware or variety store setting. The purpose of market testing is to banish speculation and fix the realities that will lead to your success.

Good market testing means good recording of data. If you run a mail-order ad, use a coupon with a response code to indicate which magazine the response originated from. For example, use a different "department number" for each publication in which you advertise, as part of the address to which the customers are to respond.

After you have "rough tuned" your marketing strategy, you can then begin "fine tuning." This involves trying different packaging, using

---

AN OUTLINE OF THE COMPANY AND COMMISSION AGENT RELATIONSHIP

The relationship between the manufacturer or "principal" and the commission sales agent generally embodies the following principles:

1. Representation is defined in a written contract between the manufacturer and the sales representative.

2. The rate of commission is defined. Any exceptional money considerations, such as a higher commission percentage or special expense allowance for new product introduction, special marketing surveys, or extraordinary detailing or technical assistance are spelled out. This might also include a provision for "advances against sales commissions."

3. Representatives are given exclusive sales rights within a specific territory, which is usually defined geographically, but occasionally, vertically by market segment. For example, a candy manufacturer might have one cadre of independent reps selling only grocery stores, and another selling drug and variety outlets. Often this arrangement is accompanied by a change in brand identification and packaging.

4. All orders, whether directly submitted by the representative or not, are commissionable.

5. If there are "house accounts" or so-called "grandfather" entanglements with prior sales reps or OEM buyers, these exceptions are noted and dealt with in the contract. (Such exceptions as house accounts are fairly common.)

6. Representatives do not carry competitive lines. They may ideally carry compatible lines.

---

different headlines in your advertising, trying different media. Find out which product features you should emphasize.

Once the product market testing is over and you've gathered data, you can make projections on a larger basis. You can use the results of your small scale sampling to estimate the general consumer response. However, you should realize that the results don't always work out the way you might expect. The public is difficult to predict. The marketplace is as sophisticated as the combination of every human want known, as capricious as every human foible, and infinitely wiser than all the "experts" of the world combined.

TARGET YOUR EFFORTS

For the fun of it, Sunday afternoons, I work in a concession stand at the local ballpark. I sell candy, soft drinks and beer. Other people pop popcorn and sell that. Sometimes we shout, "Get your fresh popcorn here!" And the south wind blows the enticing aroma down to the folks watching the game. Between innings, we go all out with our promotion and it works!

There are many other examples of "low-key" advertising. Over in the local restaurant, the tables are adorned with place mats announcing the upcoming county fair. Around the main center-spread appear calling card size advertisements for the lumber yard, the insurance man, the bank and the real estate agent. It's a small town, but virtually every businessman supports the fair, and wants to get his name in front of the public. So the printer even has to make two versions of the place mat to accommodate all the advertising.

I mention these two promotion techniques because I want to emphasize three things. First, promotion is not always the polished kind of thing you might see in a glittery television commercial. It can be very simple and down-to-earth. As long as it draws attention to your product, it is promotion.

Second, the promotion is targeted. You're not sending your message to some other planet! You're talking to people who are here and now, live prospects for your goods and services. Keep this in mind, because large-scale promotion can be vastly inefficient if the market is not targeted. It's like shouting at the sea!

Third, promotion is not necessarily expensive. Calling out "Get your fresh popcorn and soda pop here!" is not expensive. It just takes a little courage.

I'm a great believer in getting out "amongst 'em." Roll up the shirt sleeves and rub elbows with the crowd. Tell them about your product's advantages. Answer their objections. Listen to their suggestions. If you have a product you've been lucky enough to get into a retail store, hang around the store a day or two and listen and observe. If someone buys your product, walk up to them, introduce yourself and, after thanking them, try to find out what compelled them to purchase.

When Lee Iacocca, Chairman of the "New" Chrysler Corporation had put his shop in order, he decided to make himself the centerpiece of the company's promotion plan. In the television commercials, he was pictured on the assembly line, comfortable among the people putting cars together.

SELL, SELL, SELL   Mr. Iacocca exhibited pride in his company's quality and, I am told against the advice of his incredulous marketing experts, declared, "If you can find a better car, buy it!" Chrysler sales leaped, because he was talking grass roots talk, not ivory tower rhetoric. It is a classic example of excellent promotional strategy.

This example illustrates two vitally important factors in promotion. One is that you must talk straight talk. You must understand what the

TYPICAL SALES REPRESENTATIVE COMMISSION RATES

Commission rates vary greatly by industry. Sales marketing sources suggest the following range of commission percentages for various industries:

HIGH	LOW	INDUSTRY	HIGH	LOW	INDUSTRY
10.27	6.97	Abrasives	16.71	7.26	Maintenance Suppl.
15.82	8.40	Ad Specialty	13.00	7.19	Marine
11.13	5.78	Aerospace	13.40	6.76	Material Handling
15.00	6.40	Ag. Chemicals	23.21	9.75	Med. Supply & Serv.
10.27	5.83	Ag. Equipment	6.95	4.80	Metals/Processed
14.83	7.17	Architecture	6.44	3.52	Metals/Raw
8.76	4.73	Auto/Aftermarket	14.00	10.00	Mining
6.67	3.78	Auto/OEM	7.50	5.00	Mobile Homes & Acces.
15.38	7.76	Beauty/Barber	12.50	6.75	Nursery/Florist
9.21	4.36	Building Suppl.	15.00	6.30	Office/Equip & Supply
6.04	4.49	Castings/Forgings	10.40	6.60	Optical Suppies
11.78	7.20	Chemicals/Indust.	9.18	5.30	Packaging & Plastics
11.87	6.93	Chemicals/Maint.	9.85	7.45	Paints/Varnishes
13.78	8.22	Coatings	9.00	4.31	Paper Industry
14.66	9.70	Computers	14.00	5.40	Photographic
9.63	5.78	Const. Equip.	7.52	4.99	Plastics
17.49	9.21	Controls/Instrmts.	13.50	6.32	Plumbing/Heating
9.03	6.26	Electric/Consumer	16.17	7.36	Pollution Products
12.57	6.42	" /Tech & Indust.	10.11	5.62	Power Transmission
12.19	7.38	Electronic Prodcts.	16.23	6.96	Process Equipment
8.86	5.11	" /Components	17.12	8.13	Pumps
9.33	5.50	" /Consumer	7.88	5.31	Rec. Vehicles
15.14	8.07	" /Tech. Prod.	10.71	6.07	Refractories
15.68	8.94	Energy	12.29	5.65	Retail-Consumer
7.18	4.88	Fasteners	7.76	5.15	Rubber Prod.
16.75	6.25	Food/Chemicals	13.55	8.34	Safety/Security
15.53	8.38	Food/Processing	19.82	9.94	Research/Equip-Suppl.
6.67	3.33	Food/Prod. & Svc.	5.80	4.81	Screw Machine
11.88	5.91	Food/Svc. Equip.	13.00	6.55	Sporting Goods
12.14	5.70	Furniture/Furnish.	6.69	4.10	Steel Mills/Found.
15.94	6.31	Petroleum Prod.	11.14	7.43	Textile/Apparel
13.73	8.82	Government	10.75	2.38	" /Carpet-Drapery
9.47	4.67	Hardware/Hswares.	8.14	4.43	" /Industrial
9.14	6.00	H.D. Truck/Equip.	15.00	7.63	Toy/Gift/Novelty
12.38	6.70	Import/Export	7.50	6.25	Transportation
15.99	6.89	Ind/Equip-Mach.	6.45	4.17	Tubing
11.90	6.17	Ind/Supply	19.38	7.63	Utilities
11.52	7.41	Lubricants	12.50	10.00	Veterinary
5.50	3.05	Lumber Industry	16.18	8.91	Water Treatment
7.68	4.97	Machining	9.41	6.26	Welding

Generally speaking, higher commission rates are proclaimed for these situations:

1. New product introduction.
2. Temporarily to inaugurate a marketing campaign.
3. With products requiring high personal service or technical assistance in the field.
4. To provide extra compensation for retail store "detailing" where special displays must be set up or store demonstrations made.
5. To temporarily provide a "chit" or bonus compensation to sales employees of the sales representative organization.

important benefits of your product are, and tell it to them "like it is." Second, it is important how you tell them. Choose carefully among the variety of promotional techniques.

COMPETITION IS KEEN   When you promote, you're up against some truly skilled promotional people. Advertising and public relations firms sometimes spend many years working out the most compelling presentation and offer stiff competition. For that reason, you may want to avoid "downtown" and stick to the suburbs or the rural scene. That is, you might promote on a smaller scale. Consider the small town newspaper if it is a consumer product. Try a commercial on the local radio station. Limit your promotion to "point of purchase" displays. Try to get a store demonstration. Keep the promotion simple, targeted and "aim for the jugular." Try to make your message terminate in a request for an order. In short, SELL!

GET FREE PUBLICITY   One important method of promotion that many new inventors fail to consider is free publicity. Most publications will do a small editorial story on your new product, without cost! There are no guarantees that your public relations release will be picked up, and generally it takes months to get the story in print. (Magazines work months ahead, so don't be surprised by the delay.) If you have a good black and white photo of your product, send it along. The publisher wants an actual photograph or professional line drawing, not a screen print.

Sometimes, the magazine or newspaper will call and do an interview with you. You may become the subject of a "human interest" article, and be more widely read. Go for it! You never know what the next telephone call or letter will bring as a result.

In many special-interest industrial fields, there are tabloids comprised of almost nothing but publicity-generated copy. Most will give you a "freebie" to help you test the waters. Almost all such publications offer reader response cards or "bingos," as they're sometimes called. The buyer merely checks a box on a reader response card. His name is computer processed by the publisher and a compilation of leads is forwarded to you. Some reader response reports carry demographic information about the nature of the prospect's business, its size, and whether the interest in your product is immediate or long term. Most reports give the buyer's name and phone number.

CONSIDER ADVERTISING

We talked earlier about the *Ayer's Guide* as a device for assessing the market. Now, you may want to refer to this source again for selecting media to advertise your product. Another reference is *Standard Rate & Data Service*, which provides even greater detail regarding the submission of advertising and editorial copy. You'll find advertising rates and mechanical requirements in the *SRDS* guide. Again, this can be found in the business section of most larger libraries. If you want to order a copy for yourself, check the Appendix source reference at the end of this book.

Do your own field research as well. Carefully analyze the competitor's advertising techniques as well as techniques of advertisers in general. Whether you like a particular advertisement or not is beside the point. For example, many television commercials use children and children's voices to promote a product. Critics sometimes refer to this technique as the "kiddy-porn" tactic, an exploitation of both children and the viewer. But the advertising works! Almost everyone loves children, and the advertising appeals to our sentimental or, perhaps, parental instincts. If you are distracted from the set and suddenly hear a child's voice, you pay attention to the ad.

Similarly, sirens, a loud trumpet blast, or rapidly changing visual effects are used, to catch your eye and ear. It's almost like the old story about the miner whacking the mule in the forehead with a two by four. When asked why, he replied: "That's just to get his attention!" Advertising is like that. You have to get attention, then keep it. Once attention is gained, messages are aimed at character attributes such as patriotism, "nesting" or security instincts. And they are also directed to our defects of grandiosity, selfishness and false pride. It is a "no-holds-barred" business!

I have some very serious concerns about today's advertising, particularly television. I'm not alone. First, I think much advertising is sending extremely inappropriate messages. While spokesmen for such industries as pharmaceuticals will vehemently deny it, I'm convinced the drumbeat of advertising for pills and other medications imparts a strong notion among many that chemicals are quick fixes for virtually everything. Take a look at our drug problem in America.

Second, very little television advertising ever talks about product features. Wisely, they send benefit images. But they are virtually all addressed to sex, personal stature, prowess and vanity. Consumers are becoming insensitive to true value—how long the product lasts, and what it truly costs. For most consumers today, the question isn't "how much will the car cost," but rather, "what will it cost me a month" (never mind how many!) and "how important will it make me?"

What is most treacherous is that large companies can accomplish with tremendous barrages of advertising, profits which they might have better derived from truly worthy product innovation and improvements. In other words, you as an inventor are pitting the "value added" of your invention against formidable promotion campaigns that are inducing people to buy vanity rather than substance and worth. The terrible consequence is that we are becoming a nation of ignorant buyers lacking any discernment whatever about choices in the marketplace. Control of the marketplace becomes ever more locked in the hands of large companies with large promotional budgets, and that is what you as an independent inventor/entrepreneur will be up against.

THE ADVERTISING AGENCY

If you have the money, and particularly if you intend to solicit trade through the media, you should consider hiring an ad agency. The advertising specialists at an agency have the objectivity and marketing experience that often make for a successful advertising campaign. Moreover, they can save you a lot of legwork. They will cover all aspects of the advertising campaign—artwork, production, mechanicals, copy writing, media selection and more. Yet, realize that usually you pay a high fee for this talent and convenience. Many entrepreneurs cannot afford an agency, especially in the start-up phase. But don't forget, you'll be competing with those who can!

There is an alternative to hiring an ad agency, however: you can be your own advertising "contractor." Being your own contractor means that you write (and take responsibility for) your own copy, bring it to a typesetter (listed in the Yellow Pages), find a freelance artist or art studio to prepare your artwork find a photographer to take good pictures of your product, get a graphics studio to do the layout, and then find your own printer in the case of a catalog sheet. Some printers and graphics artists will advise you on successful marketing designs. But when you are your own contractor, you ultimately must rely on your own judgment in choosing advertising techniques.

CREATING SALES MATERIALS

At some point in your promotion, you may prepare a catalog sheet. Give this a lot of thought. It may become the cornerstone of your promotional effort. Identify the most important benefits of your product and make those part of your headline. Keep your message simple, straightforward and always ask for the order!

A good catalog sheet should always incorporate the basic rules of selling (as should all your promotion): introduce the product, explain benefits, overcome objections, and close the sale.

In developing a "pitch," the core of any promotion, work through scenarios with a "devil's advocate." Have someone listen to your pitch and offer all the objections they can. Then, you try to respond to those objections. Focus your message. Avoid sidetracking. Concentrate on benefits. Overcome objections.

If you have a technical invention and must include technical specifications in your copy, beware of "gobbledygook." You may understand the terms of your industry and the language of your shop, but the buyer may not. Even if the buyer is a technical person, he may have trouble understanding the relevance of certain phrases and words. So speak not only in the buyer's tongue, but in his dialect!

LISTEN TO THE MARKET

Once your advertising is out, keep in perspective the initial consumer response. Some of the positive feedback may encourage you to even loftier heights of inventive genius. Or market objections may cause you to question the invention itself. Regardless of the response you get initially, stick with what you've got, at least until you can gain a balanced interpretation of the signals.

I worked for a candy company once, and though my job was selling, I had an urge to invent new kinds of candy. Once I made a suggestion to my boss about another new kind of candy piece. He looked at me critically and said, "Bob, we make more kinds of candy than any other company in the world. Why don't you just stick to selling what we've got?" He was right! At some point, you've got to take the long gamble on what you have.

Which brings us to persistence again. Good promoters know that frequency will maximize a buying response. If you knock on enough doors and tell your story enough times, sooner or later the message will get through. It takes persistence!

I would like to be able to say that there is an easy formula for promotion. There is not. But in this chapter, I have tried to highlight the main ingredients of good promotion. Follow the proven selling formula. Tell your story to the right audience. Keep it uncomplicated. And keep doing it!

13.

Growing into a Company

Many gifted individuals, with a talent for inventing, never want to venture from the joy of their creativity. Something tugs them back to the drawing board, machine shop or laboratory, again and again. If you develop an invention to the stage where it's marketable to another company, that's fine. You can keep returning to the drawing board or laboratory. Many inventors choose that course. But if you elect to make your invention the cornerstone of a business venture —something crucial to finding investment capital—you will have to make the difficult transition from inventor to entrepreneur.

THE BOTTOM LINE

To paraphrase Calvin Coolidge: The business of business is business! And profit IS THE BOTTOM LINE! It means making a better product to serve a market. It means hard work. It means incessant attention to minute details and every fraction of a penny. It means adopting the perception that cash is just a form of capital "equipment"; it must be put to work just as your employees and machines are. In the context of your business, money is the end, but first, money must be the means to the end.

Inventors who have worked first for other companies often find it difficult to become their own bosses. The inventor who becomes self-employed often misses the "teamwork" of the larger company. He may feel lonely or have trouble establishing self-discipline outside of a large company structure. The routines of fixing hours of work, manning the phone, shifting from one kind of task to another, become burdensome.

LEADERSHIP AND PURPOSE

If the inventor hires other people to work for him, leadership and purpose are also essential. People will look to you for direction—and approval. You will find yourself face to face with problems you never thought possible, particularly people problems. As their source of work and income, the security you offer becomes paramount. Listless or insufficient instruction will translate into haphazard and shoddy workmanship. Thus, you must become the pacesetter, inspiration, father confessor and role model.

Many inexperienced managers fall into the trap of calling endless "planning" meetings at the expense of neglecting the "store." Insecure in the decision-making process, the CEO (chief executive officer) will call a meeting to discuss the purchase of bathroom tissue! This "meeting" syndrome is generally a "cop-out" from work. It is executive "goldbricking" and frequently very unproductive. In fact, because it presents a work-ethic model for plant workers or others not in management strata, it has a deleterious trickle-down effect.

Another difficulty is that management sometimes isolates itself from the realities of both its production and marketplace. There is no substitute for involvement. One cannot manage something that he or she does not understand. The only way to understand it is to roll up your sleeves and plunge in. You can understand neither your salesman's motivation nor your worker's fatigue until you do this. Not once, but again and again!

"Drift" is another problem that plagues many new managers. Days, weeks and months pass and you will discover you have made only the slightest of forward motion. Two strategies will help eliminate the affliction.

PARALLEL MANAGEMENT

First, develop the habit of parallel management: tackle as many jobs as possible simultaneously. Many people work "serially," not beginning one task until the previous one is complete. For them, work becomes a constant waiting game. Almost inevitably, they discover later that they could also have been doing something else in parallel with the first task. If a certain task isn't accomplished as hoped for, their optimism slips, and they are unequal to other accomplishments. Instead of waiting around, keep several projects going at once to eliminate boredom and improve your sense of accomplishment.

HAVE A WRITTEN PLAN

Second, have a written plan and follow it carefully. Make others in your organization aware of the plan and their role in it. Encourage teamwork and a "we" attitude. A business plan is like a financial budget. It sets goals, defines priorities and limits the range of error. Break the major plan down into its simplest components. Be realistic in making your plan! Target your objectives!

You should also try to keep your feet in the clay! Many new entrepreneurs, flushed with a modicum of early success, abandon humility altogether in exchange for a cape. Arrogance sets in, management by inspiration becomes "rule by decree." You simply cannot pound your fist on a desk and demand "esprit de corps."

SUPPLIER RELATIONSHIPS

Finally, develop sound relationships with your suppliers. You should not be content with any single source; rather have alternate and backup sources. I do not believe in an adversarial relationship with a supplier. If you chisel price, they will chisel quality. One of the best bargaining tools you have with any supplier is that of paying your bills promptly. You must also make your requirements explicit to your suppliers. If you need materials on a specific date, plan for it so that your supplier is not forced to extraordinary heroics (and pricing!) to keep your ship afloat.

SOURCES OF INFORMATION On the business text shelves of a good bookstore, there are literally hundreds of publications about managerial skills. It isn't the purpose of this book to examine all the philosophies of management, but I encourage the reader to explore some of these texts.

A HIDDEN ADVANTAGE

Japanese inventor and entrepreneur Akio Morita, head of the giant technical marvel Sony, contends there is nothing wrong with the American worker or American workmanship. Why then is America inventing, innovating and producing less manufactured goods? "Poor management," contends Morita. And the proof seems evident. Otherwise why has America transformed from the world's top producer to the world's number one creditor in just twenty short years? For that matter, why don't salesmen return telephone calls or business offices answer letters?

Therein lies a hidden advantage for the new entrepreneur. You'll be competing with firms who may be bloated with money but have grown flaccid in performance. With your high quality, prompt service and fair pricing, you'll beat them hollow. Won't you?

You may also find much practical information in publications from the special interest press. For example, if you are in the food business, you should subscribe to one or more of the magazines covering this industry. These publications keep you up to date on new equipment, and process and product developments you should be aware of. Some of the articles may even rekindle your spark of inventiveness.

In addition, most industries hold seminars dedicated to various aspects of business, from sales to production to accounting ad infinitum. These seminars are usually expensive but well worth the money—especially if they offer workshops. Likewise, most industries sponsor trade shows which preview equipment and supplies for a specific industry. You should find out about them and plan to attend. You might also consider exhibiting your own product at a trade show, if appropriate. It is an excellent "marketplace" for new items.

NETWORKING Networking is an excellent way to keep in touch, gather new information and solve business problems. A friend of mine builds church organs, and contrary to isolating himself from his competition, he maintains good relationships. He and his competition frequently trade parts, technical information and tools.

In addition, local organizations such as the Lions clubs, Chamber of Commerce and Rotary clubs, provide a useful meeting place for the exchange of views and information important to the local business climate. The local telephone directory will usually provide listings of a variety of associations and trade groups, many of which also serve a networking function.

COMMUNICATION

The key to successful management is good communication. Ultimately, your invention will grow into a strong company if you communicate with your help, your suppliers, your customers, and not the least, with your investors. Stick to what is, and avoid unrealistic projections. Above all, listen to the feedback. Listen and UNDERSTAND!

14.
Survival

Companies do not stand still. They either go forward or collapse! The seeds of self-destruction are often sown early in the game, particularly for a company founded on a single product. From the beginning, do everything you can to strengthen your company. To do that, you'll have to consider a variety of factors.

The marketplace itself offers some frightening statistics. Of all the consumer products in retail stores, 50 percent are NEW within the past five years! Second, the average life span of a consumer product in the marketplace is only five years! Third, of all the companies making products of any kind, less than 3 percent succeed if they market only one product! And finally, of all the new ventures started, less than 25 percent survive for over five years! Pretty harsh realities!

The newness of your product, one of the prime selling factors initially, will soon wear off. A lot of other hopefuls will snap at your heels, not necessarily with a like product, but with all kinds of other new products. We Americans love fads! And we thrive on change.

TECHNOLOGY ADVANCES

Automobile manufacturers are forever changing styles and adding improvements. Furniture makers vie for market "position" with ever more creative designs. Fashions change, technology advances, tastes become different, markets grow older, younger or move away altogether. Marketing and merchandising changes, with newer modes of distribution replacing older ones, and stores changing product lines.

If unaware or unprepared for all those changes, you and your company can quickly fade into oblivion. Or, and this is where your original inventive talent can serve you well, you can keep yourself not only contemporary, but in the forefront of the change. You do not have to remain a "carriage maker," but can become an automobile maker! Knowing that "they" will be nipping at your heels, you can take appropriate steps.

The trade press, and the hands-on tactic of going to the market, store or factory, becomes absolutely essential. Even unlikely kinds of publications will keep you abreast of trends.

Chain Store Age Magazine, Women's Wear Daily, Design News, Advertising Age and *Architecture* are but a few of such important publications. For example, no automobile or furniture designer worth his or her salt would try to cope without knowing what fabrics are being introduced at the Paris or Rome "openings."

UPDATE DESIGN   One of the best ways to stay ahead of the pack is to improve or update package design. For example, the pull-tape on a package of cigarettes or spout-pour plastic bottles for motor oil are both significant improvements in packaging. Especially if identified by the label "new & improved," such changes make the product more appealing to the consumer. Your product is not only what's inside, but is the total package.

WHY BUY AMERICAN?

In most cases, there is absolutely no reason a new product developer should run off to some Bangkok sweatshop to get parts for his or her product. The belief that costs can be greatly reduced is a myth! There is a hidden cost and that consists of jobs and income lost to American workers.

Everywhere in America there are excellent sources of supply for just about everything. And I've discovered there is a whole contingent of small manufacturing operations—especially in rural America—eager to supply the same articles often sourced offshore. So what if you have to pay a fraction of a cent more? What's a fraction of a cent in the context of your excellent quality, exceptional service and fair price?

A second product improvement can be the addition of a "junior" or "senior" version of the original to your product line. A jack handle can have just the wrench end for removing the lug nuts on the automotive wheel. An advanced model can also have a chisel end for prying lose the wheel cover. And together, in sufficient length and strength, they can also be the handle for pumping the jack. These may seem like obvious things. But often, the obvious is elusive, even for the inventor!

So it is important to consider how your original invention can be improved upon. If it is successful in the marketplace, others will be working on imitations. Beat them to the punch. You've invented the "Mark I"; go for the "Mark II"!

PRODUCT DEVELOPMENT   The next major survival tactic you should consider is to not just improve your original product, but to look for new items related to your original invention. For example, if it is an automotive accessory, are there other accessories waiting to be invented? If you're doing business in the air conditioning field, what do your "fresh eyes" see that others don't? This is market-driven product development.

And then, does your manufacturing capability fit the needs of products in other marketplaces. If you're making a temperature controller for the air conditioning market, might you also make a temperature controller for a special market like greenhouses? This is production-driven product development. Or you may want to purchase someone else's new invention, if it fits your manufacturing process, your marketing capability or both.

Finally, perhaps your company gains enough strength to acquire a whole different company. While this may seem an extreme departure from the "tactic" of survival, it is a means of expanding your management and other capabilities. And remember, as an entrepreneur, rather than just an inventor, your goal is profit.

SURVIVAL AND GROWTH

Companies must do more than innovate to survive. They must advertise and promote, constantly. I'm reminded of a soap company, years ago. In Baltimore, there are neighborhoods of row houses, almost all of them identical and almost all with traditional white marble steps leading to the entry. A dweller's pride (as well as his social status) was tied to the cleanliness of his marble steps.

A Baltimore company manufactured a product called "Monkey Stone Soap." Everyone bought the product, because everyone scrubbed their front steps. The company was very successful, and might still be. But they decided to quit advertising. Sales plunged, and finally, the company just went out of business. They thought they didn't have to promote. The company was not "killed" in the marketplace. It committed "suicide" at the front step!

Companies tend to stagnate. While optimism fuels the start-up, in time, it can seem that the whole endeavor is but "a tempest in a teapot." The harsh realities of the marketplace may encourage a mood of pessimism and the boat "drifts." As a friend of mine says, "The situation is characterized by obscurity." Company purpose becomes ill-defined, morale sags and sales slip. It's a contagion that at one time or another, afflicts all companies—even the giants.

As "skipper at the helm," you must provide the constant regeneration that every company needs for survival. Your team must be enthused, and it's you who must motivate them. You will come to understand that "heavy wears the crown." What do you do when the enchantment wears thin?

Years ago, I visited a customer and friend in his office. He was a tough buyer. But we liked each other, and shared similar philosophies. He had fled the concentration camps of Germany and arrived in this country virtually illiterate and penniless. I was a "bumpkin from the boonies." Within a few years, he had risen to pre-eminence in his field and made a fortune. I asked him how he did it. He said, "You've got to learn to cry when you want to laugh, and laugh when you want to cry!" I've found many occasions since to recall his words.

HAVE A DESTINATION

A trucker friend of mine likes to say, "It isn't the miles you travel, it's the distance you come." In a great sense, this is how business works. There are going to be detours and breakdowns, inevitably. The trucker and the business leader know this. So they go "on the road" prepared for the unforeseen. But they both have one thing in common. They have a destination. For the entrepreneur, it is survival and PROFIT!

15.
The Helping Industry

Over recent years, a lot of resources have come into being which serve to encourage new business development. More and more, communities are recognizing that new manufacturing businesses are necessary to revitalize their economies. And many new industries are founded on a single idea—an invention.

SBDCs In the public sector, the Small Business Development Centers (SBDCs) are providing valuable assistance to inventors seeking to gain entree to commercial success. Many of these SBDCs are attached to various educational systems, not infrequently the vocational schools or community colleges or even universities, but sometimes they stand separately, through special new business development initiatives established by various states.

The problem with the SBDCs is that many inventors or would-be entrepreneurs don't even know they exist or where. Within a given state having several SBDCs, people in one don't know others exist or what they're doing. The SBDCs are not publicized well, nor is what they offer uniform in quality or depth of service.

Another problem with the SBDCs is that they tend to become "political footballs," and both their funding and their scale of services suffer at the hand of political expediency. Too frequently, they become instruments in political objective, such as the encouragement of new industry in a preferred region of the economy, both geographical (rural versus urban or vice versa) and cultural. While this may be good, it too often leads the function and performance of the SBDC away from major realities—such as finding a market and serving it. And the hopeful inventor or entrepreneur is stranded in a web of administrative qualifications he or she simply can't service. In other words, bureaucrats do do what they do!

But there are many excellent SBDC resources and the inventor should seek them out through their state economic departments, all of which are listed in the Appendix.

There are also many private resources available. The Control Data Institutes are but one example, and throughout the country, there are many private entrepreneurial organizations, inventors clubs and shows. For example, the Redwood Falls Inventor's Congress in Minnesota has been holding annual invention exhibits for years; it is the oldest inventor's show in the country.

I do have a minor concern, however, with the value of invention shows or expositions. It may be reinforcing to exhibit an invention at a show, but if you believe this is a significant step into the marketplace, you'll probably be disappointed. At best, you may gain some public feedback, and you may even win a prize for exhibiting. But you will probably not be bringing the invention to the attention of the people you should specifically target.

Probably by far the most solid resource is the Small Business Administration which provides a wealth of help to inventors and others seeking to establish new businesses. You'll find a listing of all SBA offices and a current bibliography of SBA literature in the Appendix. SBA offices regularly conduct small seminars, and you should inquire about this at your nearest SBA office and also about the SCORE program.

USE THE LIBRARY

For some reason, inventors frequently overlook the public libraries. If you haven't been into a library in a while, you're in for a surprise! Libraries have become windows to the world, and it is positively amazing what you can learn in them. What's happened is that most libraries are now hooked into a multiplicity of computer databases and can search for literally any information available anywhere in the world. It is almost mind-boggling to discover how much even the smaller libraries have access to and can provide for you.

If you need help "searching the literature," many of the major libraries offer fee-based searches and for a fairly nominal cost, you can have a professional librarian identify a wealth of materials on almost any subject.

Again, you should view all these resources as just that—resources. You should not expect others to lay out your road map for you, tell you what direction to go, and provide the fuel for getting there. You have to provide the energy. If you seek out the assistance many of the SBDCs can offer, search the literature in your library, attend a local inventor's club (or start one!), listen to what other successful inventors are doing, you'll gain both the knowledge and excitement which can help you on your road to success.

FORGET "WELFARE ENTREPRENEURIALISM"

I take a dim view of "welfare entrepreneuralism" which is funding provided by the state for the purpose of starting a new business. More often than not, such businesses fail. Many states do provide such funding, but again, it is a circumvention of the true free enterprise system. The assumption is that state funding can make the venture go, and sometimes it does. But what really makes a venture go is that you offer the public true value, quality and service—and perhaps, just perhaps, the novelty of your invention.

I've met entrepreneurs who took the public-funded route, and then found themselves so saddled with administrative requirements their businesses could be likened to those operated by the heavy-handed management of a commune! That is no longer free enterprise. It really isn't even capitalism anymore. And despite whatever taint you may believe attached to those words, they bring greater abundance, security and happiness to more people than any other system ever devised. As an inventor, you are the cornerstone of that system. Remember this too, without it, even charity and welfare become almost impossible in our modern world.

John Kennedy paraphrased it best when he said, "Ask not what your country can do for you, but rather, what you can do for your country." But then as an inventor, you already know that, don't you?

A MODERN FABLE

Once upon a time there was a little lamb who had saved up his earnings and decided to open a butcher shop. Now Mister Lamb was a good merchant and had, as a basic philosophy of doing business, the notion that he would offer very fine quality at a very fair price and with very good service.

Mister Lamb's hunch was right! Folks liked his fine quality, good service and fair prices. And Lamb prospered. Until one day, a minor accident happened—nobody ever figured out how it came about—but Lamb sold some meat that was apparently ever so slightly tainted, and Mister Gander developed a mild indisposition for a day.

Well, some of the other animals thought the town council ought to have some way of keeping things like this from happening again, and after due consideration, the council placed Mister Pig in charge of a new bureau called the Consumer Taint Protection Bureau. Mister Lamb was somewhat irked by the necessity of this, particularly because Mister Pig insisted that Mister Lamb install certain new equipment or go out of business. Mister Lamb bought the equipment—after all, he had no choice.

Then one day, the town council decided that if they had to have a Consumer Taint Protection Bureau—which, incidentally, had now, in addition to Mister Pig, three other employees and a new office—they would require a special license fee from Mister Lamb. At the same time, they raised the town tax rate.

Inevitably, Mister Lamb had to raise his prices. But worse, Lamb's attitude began to suffer. It became more and more a chore to offer the same high quality, fair price and good service. Consequently, since he never knew when Mister Pig would ask him to make some costly new changes in his shop, or require him to keep some complicated new records, Mister Lamb began to operate his shop more and more in reaction to Mister Pig, rather than in keeping with his original philosophy.

(Cont.)

Things sagged. Mister Lamb no longer worked at obtaining the best quality. He didn't install some new labor-saving equipment he had planned on. He didn't improve his presentation and service. Lamb really didn't care anymore. And his employees got depressed and surly.

The only thing that didn't change was that folks still needed meat, and since the community was growing, they needed more meat than ever before. But because of Mister Lamb's form filling-out work load and his extra taxes (to support Mister Pig) he didn't have the money or time to improve his own shop and make it more efficient. As a result, Lamb raised his prices; partly because he just had to, and partly to pay for his annoyance.

Well, that did it. Between the extra taxes and higher cost of Mister Lamb's meat, everyone had their problems. Mister Lamb's employees had to have more money. Other shops raised their prices too, and pretty soon prices were going out of sight—and everyone was mad and blaming everyone else. Except Mister Gander. Remember him? He had the mild indisposition that started the whole thing. It turned out he hadn't gotten the indisposition from tainted meat at all. His doctor had made a mistake. So the town council met again and established another bureau, called the Bureau of Diagnostic Review, in charge of which they placed Mister Wolf.

Well, you know the rest. Except that later on, the town council got a Board of Oversight Inquiry going. And found that Mister Gander was just a bellyacher to begin with and really didn't have a problem at all. Until now. Along with everybody else. And so it goes!

16.

We've Only Just Begun

There is a legendary story about a man browsing through a yacht showroom. Eyeing one particularly splendid craft, he asked a salesman, "How much is this one?" The salesman looked at the browser indifferently and replied, "If you must ask, sir, you probably can't afford it!" So it is with inventions. If you can't think of what to invent, you're probably not an inventor. In fact, there is a great deal left to invent, and the world desperately needs inventors.

DISCOVERIES NEEDED   Let's take a look at some of the areas that cry out for new discoveries.

Robotics. There is an overwhelming scurry to develop robotic machines which can speed production in factories, worldwide. Many observers believe America has lost its competitive advantage to foreign manufacturers, most notably countries like Japan, leaders in robotics and plant automation. While much of the need falls into the category of application engineering, there are areas begging pioneer discovery.

For example, for all their dexterity, brute strength and tireless performance, robots have still not been endowed with sensory perceptions which would enable them to perform many tasks. Robot "vision" is still extremely primitive, requiring vast computer strength to interpret the feedback signals. Thus, while a robot can be easily programmed to "see" a specific object or work environment, it can become helplessly confused with an alternate object or environment.

Scientists are now working with a variety of "sight" and other sensory inputs, and computer program "templates" to improve robotic sensitivity. The robot is being made to sense normal light, but also infrared, ultraviolet, and sonic signals. Temperature, pressure, and sound sensors are being added. When the combination of signals is compared through computer templates, the robot can more "intelligently" cope with what it's "seeing."

Vast amounts of money are being plowed into robotic development. General Motors alone has recently invested $45 billion in robotics. The doors are wide open at the Pentagon to anyone who can build a vehicular robot capable of traversing unfamiliar terrain at a speed of say thirty miles per hour. And according to one worker in the field, accomplishing that task is the equivalent of inventing what nature took millions of years to do, because "mobile creatures are always getting themselves into situations."

The modernization of American factories with robots and other applied technology is critically important if we are to compete in world markets. Japanese manufacturers produce a typical automobile for $1200 less than we, largely because of their robotic lead. Steel and other manufacturing industries have virtually fled to other nations, leaving thousands in our work force jobless. Today, South Korea, West Germany, Japan, France and Canada all are advancing their worker output faster than we—much of it due to robotics.

For the individual inventor, all of this represents opportunity. Fortunes will be made by inventors who solve some of the technical problems of the industry and can apply their innovative application skills to the problems.

Medicine. Miracles are being performed in this field, and more are needed. While much of the development is extremely sophisticated, there is plenty of opportunity, because there are a lot of problems needing solutions.

It is predicted that by the year 1991, unless a breakthrough takes place, the AIDS epidemic will cost the United States $16 billion, an amount roughly equal to the medical costs from traffic accidents. But the toll is beyond cost. As with other baffling diseases like cancer, Alzheimer's disease, multiple sclerosis and even the common cold, AIDS is a dreadful and heartbreaking cost in human life.

Scientists are looking more deeply than ever into what Shakespeare so prophetically called "that mortal coil," the exquisite genetic substance DNA, which provides the blueprint for who and what we are. Anyone with a hunch is invited to join the search. Certainly, it will require training, but as truly successful inventors will confirm, imagination and creativity are seldom dimmed by increased knowledge. Rather, they are stimulated to greater heights.

Medicine has taken advantage of many recent mechanical inventions. Among them are the Jarvik artificial heart, pacemakers to regulate the heartbeat, and dialysis machines for cleansing the blood. At the forefront of this kind of work is the development of artificial limbs, which respond to neurological impulses, and implantable medication dispensers. All of these devices will undergo further improvement.

Presently, researchers are seeking ways to "soften" the Jarvik's pumping and thus lessen the possibility of blood clotting. Work is being done on improving dialysis efficiency and reducing the size and weight of the equipment. And just recently, one pacemaker manufacturer introduced an interactive device which responds to the body's variable activity and need.

In the medical field, advances are being made almost daily. In the operating rooms, in the laboratory, in diagnosis, in rehabilitation, and in all areas right up to the accounting department, there is unparalleled opportunity for the inventor.

Power. Since the Arabian oil embargo of 1973, America has become more keenly sensitive to its dependence on power. But the concern goes far beyond mere automotive or home heating fuel. We will need new sources of power for space adventure, new methods of storing power, cheaper methods of making it—indeed, new sources entirely.

Curiously, the strongest kind of power in the entire universe is also the least understood; it is gravity. If man could directly harness gravitational force, or selectively neutralize it, his energy problems would be over.

We take power for granted, which is a tribute to the industry, but there are woeful inefficiencies begging for inventive solutions. For example, high power transmission lines lose 10 to 15 percent of their power between the generator and your meter. You pay for that, of course. A coal burning boiler is only about 12 percent efficient; an atomic power plant about 3 percent efficient. Both can pollute.

Electricity produced at a river dam must be used as it is produced. If the river is flowing abundantly, the harvest of power may be abundant. If a demand for power is not present at that precise time, the water and its potential energy must simply be bypassed. How can the generator take full advantage of the water flow? By storing the electrical output in some yet undefined way, perhaps. Presently, it's shunted to other power grids which derive power from coal or nuclear fuel and, thus, can hold back their own power generation.

Expeditious power utilization presents a definite challenge. While there have already been some eloquent answers, more are needed. For example, the Tennessee Valley Authority and other hydroelectric producers sell off-peak electricity to energy-intensive factories such as aluminum manufacturers. Alternatively, surplus water is backed up into reservoirs as a means of storing potential energy. But what imaginative discovery will lead to even better ways to store this energy?

Then there is the vexing challenge of nuclear fusion. Answers continue to elude the scientists. The plain fact is that despite years of intensive effort and billions in cost, we still have apparently not yet invented fusion power. Ideas, anybody?

Environment. Great challenges face the inventor in this realm. How can we control or, even better, predict weather? How do we curb acid rain, or counter its effects? How can we reduce air pollution? How do we detoxify the growing cesspool of industrial waste? How might we improve our home and factory environments? How can we restore our seriously depleting fresh water resources? How can we reclaim our garbage? How can we more efficiently use the resources of our forests and mines?

The problems are abundant, but so are the opportunities. And inventors are providing answers. Recently, for example, an inventor, Meyer Steinberg of Upton, New York, developed a truly novel coal-burning process which scrubs acid rain-producing chemicals from the stack. He did it by slaking the coal with Portland cement! And this is just one of many such ideas in this field. Catalytic devices for vehicles, improved fuel formulations and engine design, garbage compactors and disposal units, non-polluting plastic water pipe, home well chlorinators, biodegradable chemicals and packaging are all products invented by people challenged by the need. Perhaps your name will be added to this list of inventors.

Biogenetics. Virtually no other field offers the inventor more promise or challenge than does biogenetics—with opportunities for medical innovations, potential changes in agriculture and world food supply, and financial reward. Unfortunately, as an infant industry, it is plagued with controversy and misunderstanding. Yet already, its technology is entering our world and beneficially affecting our lives. In the future, the benefits will only increase.

Biogeneticists are literally creating new species of life, living substances never before known on the planet. They've already developed microbes which eat oil slicks or cause rain to freeze and protectively coat citrus crops, bacteria which produce drugs like insulin or interferon, a crisp new celery variety, better tomatoes, hybrid vaccines and much more.

Within this broad field, amazing developments are taking place. Genetic material from nearly extinct species is being implanted in host material, and thus preserved for future generations. New super-crop varieties are being developed, by combining the disease-resistance of a "scrub" variety with the commercial qualities of another. And childless couples are parenting "test-tube" babies.

Foods with enhanced palatability or nutritional qualities are being developed in this field. Consider the fact that it is estimated that human metabolism is only 45 percent efficient, and wonder how biogenetics might improve on that.

While man has used single-celled species in fermentation and cheese-making processes for centuries, now he is learning how to biologically redesign these miniature creatures for even more useful purposes. But there are yet many questions begging answers. For example, we understand much of how photosynthesis works (70 percent of it takes place in single-celled ocean plants), but we don't have all the answers yet. They seem hidden in genetic codes. It will probably be a bio-genetic inventor who will make the discovery.

Only recently, we have discovered deep water microbes which sustain life by converting sulfurous compounds to energy, similar to the way other plants convert sunlight's energy. Imagine transferring that capability, stored in that microbe's DNA, to another microbe normally companionable with high-sulfur coal. Such is the domain of the bio-geneticist, and another set of challenges for the inventor.

Communications. This is a catchword for a vast domain which includes everything from the commercial media to personal conversation on a phone. It also includes information management of all kinds, from business accounting and correspondence to education and weapons telemetry. In fact, it is the largest sector in our national economy. It is also plagued with inefficiency begging inventive solutions.

A case in point is in the field of computers. In recent years, there have been enormous strides in the development of computer hardware. A manufacturer can now put a complete computer together in as little as seven minutes, automatically. Laser printers can now crank out computer-generated print-quality copy, dozens a minute. The result? Lower cost. But alas, programs are lacking and application of the computer to many tasks has been slowed down. There is a desperate need for programmers to bridge the gap.

IBM has developed a chip circuit so dense, it is claimed the entire *Encyclopedia Britannica* could be lodged in its memory. A dozen or more of these chips fit in a thimble! Think of the possibilities. Imagine a near-future generation of home-entertainment systems employing this technology. Consider a "boom box"-sized device capable of digitally storing hundreds of audio or video recordings. You will enter, precisely edit or erase material at will and call up day long programming as desired.

Despite all the technical improvements, the office remains the least efficient of all workplaces. How can efficiency be improved? What novel remedies will speed up matters there?

In broadcasting technology, improvements are also possible. FM broadcasting, while of better clarity, travels in a straight line, and thus has limited reach. AM signals bounce from earth to the ionosphere and travel greater distances. Might there be a way to combine the advantages of AM travel and FM clarity? Might there be another method of transmitting radio or other communication signals through water or mountains? How can jamming of a radio signal be better circumvented?

How else might we employ radio energy? In improved personal communications? In the development of personal security devices, in the navigation of autos on a highway, or in the remote operation of a farm implement, a mining machine or even a factory?

Obviously, there is limitless opportunity for invention here too.

Computers. Because this field affects every other in such important ways, it deserves a category all its own. Do not think for a moment that everything has been invented or solved here. Far from it. Circuits are being constantly improved, to a point where scientists are working virtually at a molecular level. So finely detailed is the circuit that electrons are channeled single-file through passageways. Such detailing also presents new problems. How do you keep a stream of electrons from wearing off a corner in a circuit? Why do some of the electrons mysteriously disappear from one location and show up in another? New discoveries are being made constantly about subatomic particle behavior: a realm where classical physics, chemistry and electronics all merge. Inventors here are multi-disciplined, able to apply knowledge from a variety of disciplines to solve a problem.

But still more questions need answers. For example, electronic circuits pass electrical signals at the speed of light, through a two-dimensional maze. Every task is performed serially. How can the circuit be properly manufactured in three dimensions? And then, within that cube, how can diagonal or otherwise circuitous networking be best established? Should the network resemble a circumferential or diagonal metropolitan highway? And then, how can "self-routing" and automatic redundancy be incorporated?

Other questions vex the technicians. Are there other subatomic particles besides electrons, which might be used as signal carriers? Supercooling or cryogenics has solved some of the problems but how can we further speed up the flow of electrons through the circuit? Does another time or spatial dimension exist, and does it offer other solutions? Must we go to an electro-chemical process, similar to the human brain?

The problems of speech recognition and synthesis offer a tantalizing challenge. Consider, for example, the complexity of even a simple language like English, and then imagine the enormity of recognizing Chinese, in which a single sound with different intonation can have thirty different meanings. Here, we pass over into the "software" side of computing, where virtually limitless needs exist.

Lathes, milling machines, grinders, hammer mills and almost every modern tool and machine can be operated better with computers. And for each and every need, a computer program must be written.

Given the personnel bottleneck, and the relatively high cost of developing a computer program, computers themselves are being programmed now to analyze, interpret and then write programs for other computers or machines. This is not a distant dream, but a present reality. Entire manufacturing plants are being operated by computers. And the "help wanted" signs are out everywhere.

IF THIS ISN'T ENOUGH In your everyday life, right in your home or workplace, there are inventions waiting to be invented. I'd be willing to wager that you probably have one or more in mind already. If not, here are some additional challenges:

1. Develop a process for making leather shoe soles more durable.

2. Find a way to more effectively eliminate snow and ice from walks, roadways, and rooftops.

3. Develop a single platen, multi-color printing press.

4. Invent a new musical instrument.

5. Develop a water-absorbent, flexible mold material for slip casting ceramics.

6. Perfect a process for making wall-sized, full-color holograms.

7. Invent a truly universal language.

8. Invent a tamper-proof package.

9. Develop a new structural or other material.

10. Invent a new art form or media.

11. Develop a new kind of personal transportation device.

12. Figure out how electrical power might be better distributed.

13. Invent an improved way to reclaim the desert or desalinize sea water.

14. Invent a method of suspending and restoring human life.

15. Develop a bread formula unaffected by microwaves.

16. Develop a new computer program.

17. Invent a more efficient home heating device.

18. Discover another source of illumination.

19. Invent an improved paper fastening method.

20. Develop grime-resistant paint or other coating.

21. Invent a device for harvesting electricity from a thunderstorm.

22. Perfect an improvement on washing clothes.

23. Invent a device or system for thwarting personal assault.

As the list suggests, there is virtually no limit to what might be invented. It has been said that what man can conceive, man can ultimately do. I'm not sure this is so, but it certainly illustrates the point that many of the answers lie in the questions themselves. Asking the right question is obviously the first step toward solving the problem.

If the obstacles seem formidable, don't panic. When turn-of-the-century cowboy humorist, Will Rogers, was informed of the imposing German WWI submarine threat, he said, "That's no problem. Boil the ocean!" His informant replied, "But how?" "Look," said Rogers, "I just figure out principles. You figure out the details."

Perhaps Will Rogers was right after all. We do have to figure out principles first. But details, as we have learned elsewhere, can be tackled one at a time. That's exactly how the computer, the airplane, and the automobile were developed: one detail at a time.

17.

Do You Really Have What it Takes?

At the beginning, I set out to write a book for the independent inventor. But I came to realize many other people participate in the invention business. Although most of these people function separately, together they bring new products to the marketplace. The process cannot be distinguished from the contributions and undertakings of all of the participants. Likewise, neither can these participants be excluded from a share in the invention's profits, if not its glory.

THE INVENTOR'S DESTINY

This might have been the story of the McCormick Reaper, the Apple Computer, or General Motors. All started with creative ideas. All of the individuals who fostered these companies shared a sense of destiny. All had to struggle through steps such as those outlined in these chapters. All of them were burdened with similar frustrations, doubts and apprehensions.

The primary purpose of this book was to bring some understanding of the tasks of inventing and "practicing" the invention. Likewise, it has been an objective to bring understanding of and credibility to the business of inventing. Too often, inventors are viewed as "crackpots." If you ask a scientist what he does for a living, he'll tell you he's a scientist. Not an inventor! Not even if he has a dozen successful patents will he tell you that he's an inventor. And I have never seen a classified advertisement for an inventor. Have you?

Somehow, despite the Leonardos, the Archimaedes, and the Elias Howes, the inventor seems relegated to the back pages, next to the funny paper section. Everybody chuckles at the "Rube Goldberg" characterization of the back yard inventor.

And yet, "Yankee ingenuity" built this great country. We penned in our frontier lands, and farmers twisted baling wire around baling wire to "discover" barbed wire. The Patent Office issued over a 100 patents for barbed wire alone.

100,000 PATENTS A YEAR

In 1898 the U.S. Patent and Trademark Office almost closed because they thought everything that needed to be invented, had been invented! Today, patent applications are received at the rate of 100,000 a year, are processed by a staff of over 3,000 people and over 4,500,000 U.S. patents have been issued!

What America did provide was incentive, a new continent filling with people anxious for new products. In short, it offered OPPORTUNITY! Smart people in smart companies quickly realized that survival depended on invention and innovation. Following the lead of men like Edison and George Westinghouse, smart companies began budgeting heavily into research and development. Today, companies like 3M, Texize and IBM allocate substantial budgets to new product development.

Where have the inventors gone? Many have gone to the large companies who provide the resources, the marketing expertise and the encouragement of secure financial reward. It may surprise you but companies are not granted patents. People are granted patents. And if the people work for a company funding the work, the patents are usually assigned or sold to the company as an extension of the employment agreement. Despite this, over 70 percent of all the patents granted each year are to independent inventors!

PRACTICING THE INVENTION

Given all of this, the question remains, "Do You Really Have What It Takes?" The enormity of the task ahead—which this book attempts to define—may discourage even the lionhearted. Obviously, dreaming up an idea is one thing; converting that to a tangible reality of benefit to others is a different matter entirely. And I remain skeptical about the willingness of many would-be inventors to go the full distance.

There is today within our culture too great an emphasis on novelty and gimmickry. There is too much dedication to "what's new" and little with enduring value. Perhaps we inventors are partially to blame; we after all are the folks that keep "new" alive.

But I blame much of this on planned obsolescence, the practice of purposely outmoding a product with style changes rather than substantive product improvements. And in the mad rush to make every conceivable product new again, and again, and again, too often something slips in the quality department. Then we all become targets for new products which don't work well and don't last.

No matter how novel a product is, that always comes second to quality. And no matter how dazzling your new gadget is, if you can't deliver it on time, back it with solid service and price it sensibly, you're not doing any of us any favors!

UNDERSTAND THE MARKET

What factor is most important to an inventor? Probably understanding the marketplace! Find a market and serve it. And the profits derived from a successful application of this rule prove it correct over and over again. No matter how creative or unique an invention might be, if it does not serve a market, it's destined to become little more than a tribute to self-deception!

I'm not even sure now whether inventing is truly a science. The esoteric and often elusive ingredients of creativity might better characterize the process as an "art." But unless you are fortunate enough to find someone to buy your invention outright, you must know more than the art of inventing to survive as an inventor. You have to develop an inquisitive instinct and a "why not" attitude. It must not end with the invention, but begin with it. You must also learn all you can about all the other essentials which become your handmaidens to success. While the flash of genius may take place in an isolated yoga-like trance, if the invention is to survive, it must be "practiced" in the full glare of the marketplace.

BECOME ENTREPRENEURIAL

Venture capitalists tell me their chief concern with funding inventors is that many simply do not have an entrepreneurial capability; they are not endowed with the capacity to survive in the world of commerce. Inventors tinker but they don't tally profit. A psychologist suggested to me once that tinkering with things is a far safer pursuit than negotiating with people. This may be the key. Things can be manipulated readily; people cannot.

But to be a successful entrepreneur, in addition to having a professional capacity for manipulating things, we must have the ability to manipulate people; we must induce them to buy. We must be persistent, persuasive and profit-motivated. If you don't believe that, listen to the television commercials and read the magazine advertisements.

But most of all, we must be quality-conscious. We can't invent improved products if we don't understand how existing ones work. We can't invent improved products if we don't understand why existing ones fail! We can't invent successful products if we don't understand what people want. We can't understand what people want if we don't understand people! For all of that, a lot of us will have to "smarten up," if we're to be good inventors.

VALIDATE ASSESSMENTS

Likewise, don't be too easily discouraged. Others may scoff or greet your idea with stony silence. Don't give up. Take an action step, and another, until you either validate your own assessments or confirm theirs. Stick with it. Sometimes, but not always, persistence pays off. To illustrate, there's the story of an inventor who declined failure.

Years ago, a fellow set about trying to figure out how to knot a string without letting go of both ends. He carried the string everywhere, and pulled it out often to make another try. One day, in church, he was fiddling with the string while listening to the sermon, and figured it out! His objective? Bookbinding.

Coincidentally, about that same year, the staple was invented! Did it faze him? Well, maybe a little. But he went on to make a fortune with the Elliot Addressing Machine Company. An inventor to the core! Later, another man boarded a strange craft and took it into the heavens. No, I'm not talking about Wilbur Wright. That came earlier. I'm speaking of Chuck Yeager, who took a still later "invention" on man's first supersonic flight, thrusting America into the space age.

The opportunity's still there. *Go for it!*

Appendix

I. SAMPLE INVENTION DISCLOSURE FORM

I, _____, residing at _____ street in the city of _____, state of _____ , do hereby, on this _____ day of _____, 19_____, claim to be the inventor of the following (device or process) described as follows:

(describe your invention here)

I do further attest that to my knowledge this is a new invention and that I have not seen or heard of a (device or process) like this, and the discovery is entirely my own.

As witness to this discovery and date of my disclosure, _____ , an individual residing at _____ in the city of _____, state of _____, knowledgeable in the science and art relative to my discovery, has read and claims understanding and acknowledgment of my disclosure and has hereunto also set his (her) hand and seal this date above mentioned:

_____

(Name) INVENTOR

_____

(Name) WITNESS

NOTARY

NOTE: This illustration is provided as AN EXAMPLE ONLY and you should consult your attorney for this or any other legal document.

II. SAMPLE LICENSE AND ROYALTY AGREEMENT

THIS AGREEMENT, by and between _____, hereinafter known as LICENSOR and _____, a corporation, hereinafter known as LICENSEE, WHEREAS, LICENSOR is the inventor of a novel process for manufacturing _____, and,

WHEREAS, LICENSOR is the inventor of a novel device, hereinafter known as _____, and,

WHEREAS, LICENSOR has taken the appropriate first legal steps to secure United States Letters Patents on both the aforementioned process and device, and also to secure both Trademark and Copyright registration as possible and appropriate, and,

WHEREAS, LICENSEE is desirous of the right to manufacture the aforementioned device, utilizing the aforementioned manufacturing process, and to market and sell the same, utilizing where appropriate the LICENSOR's Trademark and Copyright privileges if granted,

NOW THEREFORE, be it known that LICENSOR and LICENSEE agree as follows:

1. By consequence of the payment of One dollar ($1.00) by LICENSEE to LICENSOR, receipt of which is acknowledged; for other good and valuable consideration, and subject to the further conditions outlined hereinafter, LICENSOR hereby grants to LICENSEE the right to make and sell and to full and unrestricted use of the manufacturing process for this purpose.

2. LICENSEE agrees to pay LICENSOR a royalty in dollars equivalent to _____ percent (___%) of LICENSEE's gross profit, defined as follows:

(A) GROSS PROFIT shall be the dollar sum resulting from subtraction of contract manufacturing cost from gross sales; cash discounts if any, or selling costs shall be expressly excluded from a computation of gross profit.

3. This license is an exclusive one and LICENSOR shall grant no other license to any other party.

4. LICENSEE agrees to pay LICENSOR a minimum monthly royalty as follows: $_____ with the execution of this agreement and thereafter, $_____ each month on the first of the month.

5. A revision of the present contract manufacturing arrangement to any other manufacturing arrangement shall not be cause for impairing or lessening the percentage of royalty to the LICENSOR unless by express agreement in writing signed by both parties hereto.

6. The full cost of legal services necessary in pursuit of Patent, Trademark and Copyright protection shall be borne by the LICENSEE.

7. LICENSEE shall exercise all due diligence in safeguarding and maintaining secrecy about the aforementioned manufacturing process or other proprietary interests of the LICENSOR.

8. This license is not to be construed as an outright sale, but is granted to the LICENSEE worldwide with the exception of the Principality of Monaco which is expressly reserved to the LICENSOR.

9. This license is granted in perpetuity, shall remain in force regardless of the patentability or other proprietary legal protection available to the LICENSOR and shall be binding upon the heirs, executors and assigns of both parties.

10. Notwithstanding anything herein to the contrary, this license shall be Ipso Facto terminated in the event of bankruptcy or dissolution of the LICENSEE or upon failure of the LICENSEE to pay minimum monthly royalty payments within ten days following the first of each month hereafter.

IN WITNESS HERETO, the parties aforementioned above do affix their hands and seals this _____th day of _____, 19____.

_____

                      Inventor LICENSOR

_____

                      Company LICENSEE

_____

                             WITNESS

NOTE: This illustration is provided as AN EXAMPLE ONLY and you should consult a patent attorney familiar with current patent law and the laws of your state or jurisdiction.

III. SAMPLE NEW BUSINESS PROPOSAL

This is a proposal for the formation of a new business venture, the purpose of which is as follows:

The company will engage first in final steps in new product development; thereafter, the company will make and market a new Gymnastic device, known as the "LOVEBELL," an improved version of the traditional "dumbbell."

The unique feature of the LOVEBELL is the internal spring-loaded end weights which provide a special "live action" response from the device when used for exercise purposes.

In addition, the completed article is finished with a tough, resilient plastic coating which simulates a hand-stitched leather cover, offering an unusually attractive, scuff-resistant finish.

A United States Letters Patent, Design Patent and Trademark has been applied for by the inventor, John Jones.

PROPOSED COMPANY ORGANIZATION

The company will be organized as an incorporation under the laws of the state of Minnesota. The company name will be LOVEBELL INTERNATIONAL, INC. which has already been secured at the office of the Secretary of State.

The company will be authorized to issue 100 shares of no par value common stock, and subscription to stock ownership will be as follows:

60 shares of stock will be subscribed to by John Jones, the originator of this proposal, for which payment will be made by his cash contribution of $5,000.00, an inventory of raw materials and certain capital equipment (as specified in Exhibit A), and his other good and valuable contributions, including assignment of the pending patents for the "Lovebell" device.

40 shares of stock will be offered an interested "start-up" investor for the cash contribution of $50,000.00.

A three-member board of directors is proposed, with one membership reserved for the minority investor.

PROPOSED OPERATIONS SITE

The company will operate from rented premises costing $500.00 monthly on a one year lease arrangement, optionally renewable for four years and located at 123 Main Street, Anytown, MN. This premise provides 600 square feet of work and manufacturing space and 400 square feet of office space, deemed adequate for the company's conduct of business.

MANUFACTURING ARRANGEMENTS

Basic manufacturing will be accomplished by subcontracting two major portions of this operation.

The major cast iron metal parts will be obtained from ABC Founders, from patterns supplied by Lovebell. This supplier will furnish cast, deburred, sandblasted and acid etched parts.

The completed foundry parts will be delivered to XYZ Coated Products Company who will complete the major manufacturing step of inserting spring elements supplied by us and then molding a resilient plastic covering over the assembly; thereafter the items will be returned to our company for work completion.

When received, our personnel will apply hot stamped product identification labeling, wrap and package the finished product, and thereafter handle shipment to customers from the workspace provided.

THE MARKETPLACE

There are about 41,000 sporting goods stores throughout the United States. While not all stores carry physical exercise equipment, the majority do. This market is seen as the logical one for the "Lovebell" product.

COST AND PRICING FORMULA

Allowing for initially ample management salary and appropriate other manufacturing, administration and sales costs, the finished product cost will be under $9.50 each. The wholesale price to dealers will be initially set at $14.00 per unit, F.O.B. our plant. The consumer price is expected to be $28.00 per unit.

PROPOSED MARKETING AND SALES STRATEGY

Initially, the company intends to solicit orders from the identified sporting goods retail stores, through the use of direct mail advertising. A catalog-price sheet has been prepared and, first, a market test mailing will be made to just 1,000 of these retail stores. When successful response indicates a positive buyer interest from these stores, the remainder of the list will be purchased and the catalog sheet will then be mailed to the balance of the stores.

An introductory offer will be included in the first mail solicitation. This will consist of a printed and die-cut, life size, full color "stand-up" display of an attractive young woman hefting the new "Lovebell." This display material must yet be produced from original artwork and photographs already obtained.

A follow-up mailing will be sent to all stores not responding to the initial mailing, since a follow-up mailing usually generates a greater percentage response per promotion dollar than just a single solicitation.

Prospect names will be obtained from American List Management Company of Omaha, Nebraska.

In addition, the company will distribute a public relations "News Release" containing photos of both the product and the store display. This news release will be sent to all the major sporting goods publications nationwide, in an effort to acquaint dealers and consumers with the new product.

ANTICIPATED PROFIT FROM OPERATIONS

A 12-month financial projection is attached hereto, outlining the anticipated sales, costs and profits from the proposed operation. On the basis of this projection, a pre-tax profit of approximately $100,000.00 is estimated for the first year of operation, or about $1000.00 per share of stock, or an approximate "payout" on investment in one year.

COMPANY MANAGEMENT

John Jones, the author of this proposal, will act as Chief Executive Officer. Jones is formerly a high school athletic coach for ten years with the Anytown School, and title-winner in the inter-scholastic regional Mr. Fitness contest for the four-state region. Subsequently, Jones has operated his own public gymnasium in the city for four years. He is married, has two teenagers, and owns his own home in the city. Both he and his wife Jeannine are native Anytowners. Jones will devote full time to the company operation, having turned the operation of the gymnasium over to other supervision.

Jeannine Jones is presently employed locally as a part-time bookkeeper. She will serve the proposed company as secretary and bookkeeper on a part-time basis.

Frank Waldo is presently employed as a production supervisor in a local, non-competitive plant, and will serve part-time with the company as a supervisor of the manufacturing operations.

COMPETITIVE FACTORS

There are presently six other companies known to be manufacturing products similar to the "Lovebell." These companies are:

> 1. Red Company offering "Redbell" at $33.00 retail.
> 2. Blue Company offering "Bluebell" at $19.00 retail.
> 3. Green Company offering "Greenbell" at $29.00 retail.
> 4. Yellow Company offering "Yellowbell" at $44.00 retail.
> 5. Brown Company offering "Brownbell" at $14.00 retail.
> 6. Black Company offering "Blackbell" at $15.00 retail.

None of the competition features the "live action" characteristic of the "Lovebell." The "Redbell" and "Yellowbell" are chrome-plated articles of substantial quality. The "Bluebell" and "Greenbell" are smooth plastic-coated, sand-filled products known to rupture under rough use. "Brownbell" and "Blackbell" are both foreign imports of enameled cast iron, with fragile appearance and durability. All products are strongly competitive, well advertised and nationally promoted.

In addition to the above directly competitive products, other exercise products are available. These include spring tension devices, and the major features of the more expensive workout tables.

CAPITAL UTILIZATION

If the stock offering is fully subscribed to, a working capital fund of $55,000.00 will be available and put to the following use:

> $10,000.00 for the purchase of production molds and foundry patterns. The molds are necessary to the resilient final finishing operation in which a simulated hand-stitched appearance is created. The additional patterns are necessary for the foundry purpose.

> $5,000.00 for initial foundry and molding contract manufacturing costs.

> $5,000.00 for the purchase of a 5 ton Harbro hot-stamping press and appropriate dies and supplies for affixing product identification labels to the "Lovebell."

$5,000.00 for packaging, shipping room supplies and miscellaneous shop expenses.

$10.000.00 for the purchase of display and advertising catalog sheets, mailing lists, postage and labor in the company's mail-order promotion efforts.

$10,000.00 for labor, office expense and miscellaneous supply.

$10,000.00 reserve for unanticipated needs.

PRODUCT LINE ENLARGEMENT

The initial "Lovebell" model is a 10 pound version, deemed the most popular of sizes. However, it is intended to broaden the line to include a 5 pound and 20 pound version.

With the establishment of the "Lovebell" brand name and a dealer marketing program, additional physical fitness products may be introduced. These would include running "weights," also with the distinctive molded leather appearance, as well as other categories of exercise devices.

CONCLUSION

Physical fitness is of growing interest to people in all walks of life. Its virtue is constantly promoted by every source from the President's Council on Physical Fitness to the American Medical Association. The interest in fundamental exercise devices, as well as very expensive ones, is translating into a fast growing market. The value of physical fitness is so widely acknowledged that there is virtually no foreseeable limit to business opportunity derived from of it. And Lovebell International, Inc. can become a vital part of this opportunity.

FOR FURTHER INFORMATION, CONTACT:

John Jones
123 Main Street
Anytown, Minnesota 55345

IV. SAMPLE COST ACCOUNTING

Accounting for manufacturing is sometimes overwhelming if you are not used to performing the tedious mathematical functions necessary to plot your course. A personal computer is a great help, but watch out! They do not leave good "audit trails," and a simple formula mistake can be disastrous!

The first three pages are EXAMPLE spreadsheets illustrating how an operation is budgeted and then what happens with different scenarios.

1. EXAMPLE A shows a budget arrived at here based on the sale of 1,000 units at $38.00 each. Various manufacturing and overhead costs are estimated and, under the budget, a pretax profit of $12,389.31 is estimated on sales of $38,000.00.

> Column 2 (% of Sales) reflects a percentage calculation for each item within the budget. Thus, labor is found to have a .2197 percentage, nearly 1/4 of the sales price.

> Net profit under column 2 comes out to .2596; almost 26%.

> Column 3 shows an actual monthly performance. Note how reality departs from the projected. And column 3 shows how this reality works out in percentages. Here, net profit actually turned out to be .2260 (about 23%), a drop of about 3% from the budget estimate.

> Column 4 depicts the dollars and cents values for each item of the actual performance reflected in column 3. In this case, net profit came to $8.5867 (or almost $8.59) for each item made and sold during the period.

2. In EXAMPLE B the same figures have been manipulated to reflect a *decline in production and sales to 60%* of the budgeted target. Here, we find we're at almost the "Break-even" point. The company made just $78.74 total net profit for the month, or almost $0.13 on each of 600 items made and sold.

3. In EXAMPLE C, manufacturing and sales have been *increased to 120% of the budget.* Note that profit jumps substantially to 38% of gross sales!

> For simplicity, none of these illustrations show what happens when certain of the cost factors are changed. For example, an increase or decrease in factory or overhead costs would alter the results in a way similar to the way an increase or decrease in units of production does.

4. EXAMPLE D portrays a slightly different spreadsheet in which a 4 month period is analyzed. Here, all columns are carried out to the 4th decimal place. (In the first 3 illustrations, only columns 2, 4 & 5 were).

> An additional feature of this final illustration is that of *cash flow.* Cash flow calculations are extremely important in operating a business, and Venture Capital companies will generally ask for this kind of accounting projection.

NOTE: These accounting examples are by no means complete. They are presented merely to illustrate the kind of financial accounting which manufacturing operations employ. *You should consult with your accountant for advice on establishing a system which will suit your needs.*

XYZ ENTERPRISES – PROFIT & LOSS STATEMENT
(Example A)

MONTH	BUDGET	% of SALES	ACTUAL	% of SALES	PER UNIT
GROSS SALES	38000.00	100.0000	32300.00	100.0000	38.0000
Units	1000.00	100.0000	850.00	100.0000	38.0000
MANUFACT. COSTS					
Plant Rent	500.00	0.0132	500.00	0.0155	0.5882
Utilities	120.00	0.0032	98.50	0.0030	0.1159
Insur. - Liab.	100.00	0.0026	100.00	0.0031	0.1176
Labor	8347.00	0.2197	7920.00	0.2452	9.3176
W/Comp. Ins.	333.88	0.0088	316.80	0.0098	0.3727
F.I.C.A	500.82	0.0132	475.20	0.0147	0.5591
Raw Materials	6345.19	0.1670	5448.00	0.1687	6.4094
Equip. Amort.	1103.45	0.0290	1103.40	0.0342	1.2981
R & D	564.00	0.0148	389.67	0.0121	0.4584
Freight	756.00	0.0199	547.85	0.0170	0.6445
Misc. Fact. Exp.	452.35	0.0119	38.56	0.0012	0.0454
MFG. COSTS:	19122.69	0.5032	16937.98	0.5244	19.9270
MANUF. PROFIT:	18877.31	0.4968	15362.02	0.4756	18.0730
SALES & ADMIN. EXP					
Office Exp	175.00	0.0046	175.00	0.0054	0.0454
Telephone	248.00	0.0065	265.00	0.0082	0.3118
5% Sales Com.	1900.00	0.0500	1615.00	0.0500	1.9000
Salaries	2500.00	0.0658	2500.00	0.0774	2.9412
Automobile	100.00	0.0026	85.00	0.0026	0.1000
Adv. & Prom.	1250.00	0.0329	1190.00	0.0368	1.4000
Legal	175.00	0.0046	250.00	0.0077	0.2941
Accounting	100.00	0.0026	125.00	0.0039	0.1471
Misc. Exp	100.00	0.0026	33.60	0.0010	0.0395
OVERHEAD:	6548.00	0.1723	6238.60	0.1931	7.3395
Pre-Tax Profit:	12329.31	0.3245	9123.42	0.2825	10.7334
Less Est. Taxes	2465.86	0.0649	1824.68	0.0565	2.1467
NET PROFIT OR LOSS:	9863.45	0.2596	7298.74	.2260	8.5867

XYZ ENTERPRISES — PROFIT & LOSS STATEMENT
(Example B)

MONTH	BUDGET	% of SALES	60% of: BUDGET	% of SALES	PER UNIT
GROSS SALES	38000.00	100.0000	22800.00	100.0000	38.0000
Units	1000.00	100.0000	600.00	100.0000	38.0000
MANUFACT. COSTS					
Plant Rent	500.00	0.0132	500.00	0.0219	0.8333
Utilities	120.00	0.0032	98.50	0.0043	0.1642
Insur. - Liab.	100.00	0.0026	00.00	0.0044	0.1667
Labor	8347.00	0.2197	7920.00	0.3474	13.2000
W/Comp. Ins.	333.88	0.0088	316.80	0.0139	0.5280
F.I.C.A	500.82	0.0132	475.20	0.0208	0.7920
Raw Materials	6345.19	0.1670	5448.00	0.2389	9.0800
Equip. Amort.	1103.45	0.0290	1103.40	0.0484	1.8390
R & D	564.00	0.0148	389.67	0.0171	0.6495
Freight	756.00	0.0199	547.85	0.0240	0.9131
Misc. Fact. Exp.	452.35	0.0119	38.56	0.0017	0.0643
MFG. COSTS:	19122.69	0.5032	16937.98	0.7429	28.2300
MANUF. PROFIT:	18877.31	0.4968	5862.02	0.2571	9.7700
SALES & ADMIN. EXP					
Office Exp	175.00	0.0046	175.00	0.0077	0.0643
Telephone	248.00	0.0065	265.00	0.0116	0.4417
5% Sales Com.	1900.00	0.0500	1140.00	0.0500	1.9000
Salaries	2500.00	0.0658	2500.00	0.1096	4.1667
Automobile	100.00	0.0026	85.00	0.0037	0.1417
Adv. & Prom.	1250.00	0.0329	1190.00	0.0522	1.9833
Legal	175.00	0.0046	250.00	0.0110	0.4167
Accounting	100.00	0.0026	125.00	0.0055	0.2083
Misc. Exp	100.00	0.0026	33.60	0.0015	0.0560
OVERHEAD:	6548.00	0.1723	5763.60	0.2528	9.6060
Pre-Tax Profit:	12329.31	0.3245	98.42	0.0043	0.1640
Less Est. Taxes	2465.86	0.0649	19.68	0.0009	0.0328
NET PROFIT OR LOSS:	9863.45	0.2596	78.74	0.0035	0.1312

XYZ ENTERPRISES – PROFIT & LOSS STATEMENT
(Example C)

MONTH	BUDGET	% of SALES	120% of: BUDGET	% of SALES	PER UNIT
GROSS SALES	38000.00	100.0000	45600.00	100.0000	38.0000
Units	1000.00	100.0000	1200.00	100.0000	38.0000
MANUFACT. COSTS					
Plant Rent	500.00	0.0132	500.00	0.0110	0.4167
Utilities	120.00	0.0032	98.50	0.0022	0.0821
Insur. - Liab.	100.00	0.0026	100.00	0.0022	0.0833
Labor	8347.00	0.2197	7920.00	0.1737	6.6000
W/Comp. Ins.	333.88	0.0088	316.80	0.0069	0.2640
F.I.C.A	500.82	0.0132	475.20	0.0104	0.3960
Raw Materials	6345.19	0.1670	5448.00	0.1195	4.5400
Equip. Amort.	1103.45	0.0290	1103.40	0.0242	0.9195
R & D	564.00	0.0148	389.67	0.0085	0.3247
Freight	756.00	0.0199	547.85	0.0120	0.4565
Misc. Fact. Exp.	452.35	0.0119	38.56	0.0008	0.0321
MFG. COSTS:	19122.69	0.5032	16937.98	0.3714	14.1150
MANUF. PROFIT:	18877.31	0.4968	28662.02	0.6286	23.8850
SALES & ADMIN. EXP					
Office Exp	175.00	0.0046	175.00	0.0038	0.0321
Telephone	248.00	0.0065	265.00	0.0058	0.2208
5% Sales Com.	1900.00	0.0500	2280.00	0.0500	1.9000
Salaries	2500.00	0.0658	2500.00	0.0548	2.0833
Automobile	100.00	0.0026	85.00	0.0019	0.0708
Adv. & Prom.	1250.00	0.0329	190.00	0.0261	0.9917
Legal	175.00	0.0046	250.00	0.0055	0.2083
Accounting	100.00	0.0026	125.00	0.0027	0.1042
Misc. Exp	100.00	0.0026	33.60	0.0007	0.0280
OVERHEAD:	6548.00	0.1723	6903.60	0.1514	5.7530
Pre-Tax Profit:	12329.31	0.3245	21758.42	0.4772	18.1320
Less Est. Taxes	2465.86	0.0649	4351.68	0.0954	3.6264
NET PROFIT OR LOSS:	9863.45	0.2596	17406.74	0.3817	14.5056

XYZ ENTERPRISES — PROFIT & LOSS STATEMENT
(Example D)

MONTH:	January	February	March	April
MANUFACTURING REVENUE				
Goods Sold	1800.00	943.00	1235.00	1564.00
Other income	110.00	200.00	145.00	190.00
TOTAL INCOME	1910.00	1143.00	1380.00	1754.00
MANUFACTURING EXPENSES				
Rent	220.00	220.00	220.00	220.00
Telephone	45.00	36.00	34.00	45.00
Electricity	52.00	54.00	44.00	76.50
Automobile	56.70	67.00	87.40	34.20
Printing	15.00	165.00	65.00	33.25
Office Supply	23.40	44.00	19.00	12.54
Plastic	459.00	238.40	165.00	220.00
Total Expenses	871.10	824.40	634.40	641.49
PROFIT OR LOSS	1038.90	318.60	745.60	1112.51

--------------------PERCENT OF TOTAL INCOME--------------------

	January	February	March	April
Rent	0.02	0.03	0.02	0.03
Telephone	0.03	0.05	0.03	0.04
Electricity	0.03	0.06	0.06	0.02
Automobile	0.03	0.06	0.06	0.02
Printing	0.01	0.14	0.05	0.02
Office Supply	0.01	0.04	0.01	0.01
Plastic	0.24	0.21	0.12	0.13
Total Expenses	0.46	0.72	0.46	0.37
PROFIT OR LOSS	0.54	0.28	0.54	0.63

------------------------------CASH FLOW------------------------------

	January	February	March	April
Beginning Cash	100.00	1138.90	1447.50	2203.10
Profit for Month	1038.90	318.60	745.60	1112.51
Ending Cash	1138.90	1457.50	2193.10	3315.61
% Change for Month	11.39	1.28	1.52	1.50
Cumulative % Change	11.39	14.58	21.93	33.16

--------------------------COST ACCOUNTING--------------------------

	January	February	March	April
Units Sold	60.00	30.00	40.00	47.00
Selling Price Each	31.83	38.10	34.50	37.32
Cost Per Unit	14.52	27.48	15.86	13.65
PROFIT PER UNIT	17.31	10.62	18.64	23.67

V. GLOSSARY

ALLOY—A composition of metal.

AMORTIZATION—An accounting procedure for apportioning the cost of capital equipment costs over a period of time, generally the useful life span of the equipment.

ANGEL—A private investor who provides seed capital for a new business venture.

ASSETS—The tangible properties of a person or company.

BANKRUPTCY—A court-directed financial termination or reorganization of a company, generally by consequence of an inability to discharge the company obligations.

BLOW MOLDING—A hollow, extruded plastic tube is positioned so mold cavities close, while air is forced simultaneously into the tube, forcing it to take the shape of the mold; thereafter, the molds are opened and the finished part ejected. Used in making plastic bottles, toys and other shapes.

BOND—A corporate note. An unsecured and interest-bearing obligation of the corporation to an individual or company. The holder of a bond is not an equity investor, but rather, a creditor. If the bond is due at a greatly extended period of time, it is a long-term bond and the holder is a long term creditor.

CAPITAL EQUIPMENT—The manufacturing or production equipment of a company, sometimes including buildings and land or 'real' property.

COLLATERAL—Goods or property used to secure a loan of money with. A loan which is 'collateralized' is said to be secured by the pledge of property known as collateral. Also known as a 'chattel.'

COPYRIGHT—The registration by the Library of Congress of an original work of art, or a copy of the original. Issued by classes, including printed manuscripts, published articles from original manuscripts, original three dimensional works of art and replications of original works, musical compositions, computer programs and others. The term of a Copyright is for the life of the author plus 50 years, but varies under some circumstances.

COST ACCOUNTING—A particular form of accounting in which units of production are divided into dollar amounts to arrive at a 'per unit cost.' In an alternate form, a percentage figure is derived from the division of a dollar amount by another dollar amount; the latter is generally expressed to the 4th or greater decimal equivalent.

DEBENTURE—A corporate 'note' of obligation, often issued as a 'convertible debenture,' meaning that it guarantees a payment of interest while outstanding, or it may be converted by a pre-defined formula to stock ownership in the company.

DESIGN PATENT—Also issued by the Patent Office and pertaining to a 'new, original and ornamental design for an article of commerce.' A design patent protects only the appearance of an article and not its structure or utilitarian features. The term of a Design Patent is 14 years, and is extendible under certain conditions.

E.D.M. MACHINING—A form of machining employing direct electrical current to 'arc' away metal. 'Electric Discharge Machining.'

ENTREPRENEUR—A businessperson; one who starts and operates a business enterprise.

EXTRUSION MOLDING—Continuous formation of tubes, pipes and shapes of metal or plastic through special forming dies.

F.I.C.A—The Social Security contribution of both employee and employer.

F.O.B—A shipping term, meaning 'Freight on Board,' and in trade, designates to what point freight charges are passed on to the seller. Thus, 'F.O.B.: our warehouse' means the customer pays for the transportation from your warehouse to his.

FIRST ARTICLE APPROVAL—Inspection and approval of the first production quality output from a manufacturing process. A quality control function.

FISCAL—Having to do with the financial side of an endeavor. 'Fiscal accounting' is financial accounting; the bookkeeping of a venture or enterprise.

HOBBING—A now nearly obsolete metal working process. A ductile metal billet is placed within a hardened retainer ring and then, with extreme pressure, a hardened steel pattern is plunged into the metal to form a cavity. This process has largely been replaced by 'E.D.M' machining. A variation of the technique, employing soft clay, is still used by sculptors and machinists to 'preview' the accuracy of die cavities.

INJECTION MOLDING—Applicable to both plastic and metal. The process of extruding molten material into mold cavities, allowing the hot material to cool and removing the molded object thereafter.

INTANGIBLE ASSETS—Those values held by a person or company which have to do with credibility, customer and other good will, the trade secrets, management capability, marketing skills and professionalism of the individual or organization.

INVESTMENT CASTING—A metal casting process. A wax model of the part is prepared and then embedded in a special slurry of refractory 'slip' ('investing'). After the slurry has been poured, it is vacuumed to remove air bubbles. When the 'slip' has set up and dried, it is placed in a high temperature 'burnout' oven, the material becomes 'fused' and the wax melts or burns off, leaving a hollow void. Thereafter, molten metal is poured into the 'cavity' remaining, often by spin casting. Used in the manufacture of highly intricate parts such as jewelry and sculpture or pieces with undercuts.

IPO—An initial public offering of a company's stock.

LAYUP—A term used to describe the plastic manufacturing process involving the use of an original pattern upon which a fiber cloth or mat, such as of fiberglass, burlap or sisal, is combined with catalyzed polyester or epoxy liquid resin to form, when hardened, a variety of shapes from automobile bodies to recreational vehicles to sculpture and football helmets.

LETTERS PATENT—A federal government grant issued by the Patent Office conferring a right to an original inventor to 'exclude others from making and selling' an invention. A Letters Patent must define a 'new and useful process, machine, manufacture, or composition of matter.' The term of a Letters Patent is 17 years, and is extendible under certain conditions.

LIABILITIES—The obligations of a company.

MACHINING—A manufacturing method employing milling machines, either vertical or horizontal, which cut away metal, drill holes or slots and shapes, tap threads and otherwise 'sculpt' solid material, generally wood, metal or plastic.

MARKETING—The art of understanding what causes people to buy.

PATTERN—A working original object used as a guide or template for the creation of a final mold. Employed in 'contour' machining and in sand casting metal parts. Also used in thermoforming plastics.

PROCESS—An endeavor involving the conversion of materials or components from one form to another, generally raw materials into finished goods. To process is to manufacture.

PROTOTYPE—A working model. Also, (v.) to prototype. A handmade sample of a new article; a working model of an invention.

PUBLIC DOMAIN—Belonging to the general public; as 'in the public domain.' An invention must be submitted for patent protection within a period of time or it passes into the public domain.

QC—'Quality Control.' Measures taken to assure accuracy and continuity of production parts or process elements

RESIN—Any of a number of plastic materials.

ROLLOVER—Converting a start-up company to a different form, such as a publicly funded company. Also, a private buy-back of a public company. Also, a merger, or bankruptcy.

ROTATIONAL MOLDING—Also known as 'Slush Molding.' The process involving the placement of plastic powder into a mold cavity, then rotating and heating the mold to cause the powder to melt, after which the mold is cooled and the piece removed. Used in making hollow parts such as large plastic tanks and small hollow figurines in metal. Sometimes used with pre-polymerized liquid plastic resins which set up into hollow objects.

RTV—'Room Temperature Vulcanizing.' A material, usually a silicone or urethane plastic which, when mixed with a suitable catalyst, sets up at room temperature to a rubberlike solid. Used extensively in making replicates of original models or prototype parts.

SAND CASTING—A metal forming process. A cavity is formed by pressing a pattern into a bed of sand. Molten metal is poured into the cavity and cools to harden. Generally employed for rougher and sturdier parts such as manhole covers, machinery bases and the like.

SBDC—'Small Business Development Center.' An organization, usually sponsored by the state, and often associated with the educational establishment such as university, community colleges or vocational training schools and dedicated to assisting inventors and entrepreneurs in establishing new businesses.

SBIC—'Small Business Investment Company.' Venture capital companies funded through the Small Business Administration.

SEC—The Securities and Exchange Commission, a federal regulatory agency which defines the rules governing the public issuance of corporate stock.

SEED CAPITAL—Funds provided to start up a business, sometimes by the entrepreneur, but often by an outside investor, such as an 'Angel' or a Small Business Investment Company (SBIC) or venture capitalist.

SPIN CASTING—A metal casting process in which molten metal is centrifuged into the mold cavities, rather than poured in. Used in making high finish and accurate parts.

SPREADSHEET—A work form in accounting, generally a grid into which labels and numbered information are placed to form ledgers and journals of accounts.

STOCK—A legal instrument used to designate the ownership of shares of interest in a corporation and evidenced by a 'stock certificate.' Varies from state to state, and is sold by classes, such as common, preferred, preferred interest bearing, etc. A stockholder is a corporate owner, in whole or part.

STOCKHOLDER EQUITY—The ownership interest in a corporate business.

SWEGING—Affixing two pieces together with a post located on one part and passing through another; thereafter the extending post is peened or flattened over, similar to riveting.

TANGIBLE ASSETS—The real and material assets of a person or company.

THERMOFORMING—Sometimes referred to as 'vacuum' forming. A sheet of plastic is heated and then drawn or pressured down over a pattern to create a finished shape. Used in the manufacture of everything from small plastic jelly cups to boats.

THIN CORPORATION—A corporation which is organized with a minimum of equity investment and must then borrow money through the issuance of bonds, warrants, debentures or other 'instruments' in order to acquire needed working capital.

TRADE STYLE—The image and characterization of a business endeavor. A business operates as a franchise, a retailer, a manufacturer, or a mail order house, for example.

TRADEMARK—A grant issued by the Patent Office giving the owner the right to 'exclude others from the use of a "Mark" in commerce.' The Trademark is issued by class, such as food, agriculture, machinery, and the like, and limited to the class. The Trademark may be valid for so long as it remains in interstate commerce. Also issued by states.

VENTURE CAPITAL—Funds invested by a 'venture capitalist,' usually an 'angel' or a small business investment firm. Such funds are generally characterized as 'equity' capital or more specifically, 'risk' funding. Venture capital is often 'start-up' investment capital for a new company.

WORKING CAPITAL—The funds used for the daily financial working needs of the company. Funds with which to purchase raw materials, pay salaries and expenses, and to finance or 'carry' accounts receivables on the books of the company. Usually distinguished from the capital equipment funding, and generally 'liquid' in the sense that the funds rotate through cycles from cash to inventory to sales to cash again within a short time span.

VI. ADDITIONAL SOURCES OF INFORMATION

Aldrich Scientific
Box 675
Helotes, TX 78023

Offers $3.00, 225 page catalog of glassware, chemicals, laboratory and biological supply items.

American Business Lists, Inc.
P.O. Box 27347
Omaha, NE 68127

A reliable source of mailing lists of the 'compiled' kind.

Boardroom Reports
330 West 42nd Street
New York, NY 10036

Publishes excellent business periodicals and texts including *Boardroom Reports, Bottom Line/Personal, Tax Hotline, Boardroom Books, Breakthrough, Priviledged Information* and *Health Confidential*. Phone (212) 239-9000

Business & Professional Mailing Lists—Catalog
Zeller & Letica, Inc.
15 East 26th Street
New York, NY 10010

A major list provider offering over 10,000 different mailing lists, both compiled and qualified. Excellent for planning a marketing strategy.

CFO, The magazine for Chief Financial Officers
CFO Publishing Company
P.O. Box 844
Farmingdale, NY 11737-9944

A very good publication for financial officers in company organizations. Very good information about accounting, particularly the 'frontier' computer variety, arranging financials for public offerings, and more.

Commissioner of Patents and Trademarks
Patent & Trademark Office
Washington, DC 20231

Provides information about registered Patent Attorneys nationwide and about requirements for the submission of applications for patents, and trademarks.

Consumer Product Safety Commission
Washington, DC 20207

Your source for information on product safety requirements of all kinds.

Copyright Office
Library of Congress
Washington, DC 20559

Provides bulletins and forms for registering Copyrights.

Creativity—A. Dale Timpe
Kend Publishing
ISBN 0-8160-1463-9

An extensive examination of how to carry creativity over into effective business management.

Directory of Conventions
Successful Meetings
633 Third Avenue
New York, NY 10017

The most comprehensive directory of meetings and conventions in the U.S. Annual with mid-year supplement. $105. For new product developers planning to attend or exhibit, this is the definitive guide to trade shows, exhibits, and conventions. Track industry events, locate supply sources, and make important business contacts. Phone (800) 624-6283 to order, or find it in your public library.

Discover Magazine
Time-Life Bldg.
New York, NY 10020

A fine publication with wide-ranging coverage of a variety of scientific news, worldwide. Well illustrated articles, and thought provoking material. A stimulant for the inventive mind!

Electronic Industry Telephone Directory
Harris Publishing Company
2057 Aurora Road
Twinsburg, OH 44087-1999

The most concise directory of the electronics industry published.

Entrepreneur Magazine
2311 Pontius Avenue
Los Angeles, CA 90064

A very good magazine covering a wide range of entrepreneurial interests, from new products and companies to manufacturing and retailing.

Epoxy Plastic Tooling Manual.
Ren Plastics, a Ciba-Geigy Company
5656 South Cedar
Lansing, MI 48909

A very good manual for the inventor or prototyper. Well written and illustrated to show model making and production techniques.

Game Inventors Of America, Inc.
Box 58711
World Trade Center
Dallas, TX 75258

A member funded society sharing information and experience. Send for their free brochure.

GAMEPLAN: The Game Inventor's Handbook —
Stephen Peek
Betterway Publications, Inc.
P.O. Box 219
Crozet, VA 22932

The reference for creating, developing, manufacturing, and marketing of games.

Foredom Miniature Power Tools — Catalog
Foredom Electric Co., Inc.
A Division of Blackstone Industries, Inc.
Bethel, CT 06801

A useful catalog of tools and supplies for carving. (Also check your local wood carving supply store.) Illustrates the company's products. A helpful reference and source for the inventor and prototyper.

Hagenow Laboratories
1302 Washington
Manitowoc, Wl 54220

Offers a $1.00 catalog of laboratory chemicals and supplies.

Hexcel Corporation — Catalogs
P.O. Box 2197
20701 Nordhoff Street
Chatsworth, CA 91311

A major supplier of epoxy and urethane resins for the model maker and prototyper.

Home & Auto Magazine
Harcourt Brace Jovanovich Publications
747 Third Avenue
New York, NY 10017

A leading publication covering the automotive aftermarket industry, including also lawn & garden, motorcycles, RV's, snowmobiles, scooters and wheel goods.

Homemade Money — Barbara Brabec
Betterway Publications, Inc.
P.O. Box 219
Crozet, VA 22932

A splendid A to Z reference for the beginning entrepreneur and an excellent refresher for even the seasoned small business person.

HP State Manufacturers Directories
Harris Publishing Company
2057 Aurora Road
Twinsburg, OH 44087-1999

Compiles and publishes or distributes directories of manufacturers in over 3,500 classifications for all states, Canada and Europe. Prices range from $29.00 to $179.00 per state or geographical directory. Write for availability and pricing. Excellent for finding new markets or supply sources. Also provides diskette output, mailing labels, etc.

Hysol Division,
The Dexter Corporation
15051 E. Don Juliard Road
Industry, CA 91749

A good source for a wide variety of prototyping materials, including metal powder-filled casting resins, laminar lay-up and other materials useful to the prototyper or model maker. Hysol products are usually locally available.

IASCO—Catalog
Industrial Arts Supply Company
5274 West 36th Street
Minneapolis, MN 55416

Supplies schools and other institutions with a wide variety of materials and supplies useful to the inventor or prototyper.

Illustrated Fact Book of Science
Arco Publishing Co.
215 Park Avenue S.
New York, NY 10003

A very fine book, excellently illustrated, covering the whole length and breadth of science. Stimulating!

IMS Ayer Directory of Publications
IMS Press Inc.
426 Pennsylvania Avenue
Fort Washington, PA 19034

An excellent reference to the publishing world, with particular emphasis on the editorial content. Gives publishers' and editors' names. A useful companion to SRDS, and a good place to begin a publicity campaign.

In Business Magazine
JG Press, Inc.
Box 323
Emmaus, PA 18049

A stimulating magazine, available at newsstands, covering matters of interest to small business.

Inc Magazine
Inc Publishing Corp.
P.O. Box 2538
Boulder, CO 80322

A fine publication with strong emphasis on the people side of the business scene, from start-ups to 'Blue Chip.'

Inventor's Resource Directory
Consumer Products Edition
Birchmont Press
P.O. Box 489
Four Oaks, NC 27524

An inventor's guide to funding sources, retail marketplaces and new product licensing prospects.

Inventor's Workshop International Education Foundation—IWIEF
Inventor's Workshop International
3201 Corte Malpaso, #304
Camarillo, CA 93010

A multifaceted organization offering a variety of services. Publishes *Invent! Magazine*, offers patent & trademark searches through PatentSaver, invention feasibility analyses through the Prep program, and licensing and technology transfer services through its subsidiary Mindsight Corporation. Also provides books and other services at discount to members, and sponsors local inventor chapters in various cities. Write for current catalog of services and membership costs.

Licensing Industry Merchandisers' Association
200 Park Avenue
Suite 303E
New York, NY 10166

Provides licensing guideline information.

Lucite Acrylic Resins Design Handbook
E. I. du Pont de Nemours & Co, Inc.
Wilmington, DE 19898

An excellent handbook for the designer. Good basic study for the inventor and prototyper. Comprehensive information on acrylics.

Managing Automation Magazine
Thomas Publishing Company
11 Penn Plaza
New York, NY 10001

A 'controlled circulation' monthly for people working in automation of all kinds, from CAD/CAM design to robotics.

Mark's Standard Handbook of Mechanical Engineering
McGraw Hill Book Company
1221 Avenue of the Americas
New York, NY 10020

A monumental compendium of information on virtually every aspect of mechanical engineering. Expensive, but a must for the mechanical inventor or design engineer.

Materials Science and Metallurgy (Pollack)
Reston Publishing, Inc.
A Prentice-Hall Company
Reston, VA 22090

A fine overview of various material specifications to acquaint the inventor with metal alloys, plastics, wood and other substances. A recommended reference.

Merrell Scientific
1965 Buffalo Road
Rochester, NY 14626

This firm offers a $2.00 catalog of chemicals, glassware, science and hobby equipment of interest to the inventor or prototyper.

National Institute of Standards & Technology (NIST)
U.S. Department of Commerce
Gaithersburg, MD 20899

This is the new name for the former 'National Bureau of Standards.' NIST is a federal agency whose purpose is 'to assist industry in the development of technology and procedures needed to improve quality, to modernize manufacturing processes, to ensure product reliability, manufacturability, functionality, and cost-effectiveness, and to facilitate the more rapid commercialization ... of products based on new scientific discoveries.' Write for current catalog materials and information.

National Technical Information Service (NTIS)
U.S. Department of Commerce
5285 Port Royal Road
Springfield, VA 22161

This is the central federal source for government scientific and technical information. Through NTIS, you can access engineering and research information being conducted by over 350 federal agencies, U.S. industry and university contractors, over 20 foreign nations, and enormous database resources, technology catalogs, government inventions available, and extensive references of all kinds. Send for current NTIS Products & Services Catalog for a full listing.

NEW CATALOG
Post Office Box 37000
Washington, DC 20013

Offers a free catalog of nearly 1,000 U.S. government bulletins on virtually every subject.

Plastics Mold Engineering Handbook (Dubois & Pribble)
Van Nostrand Reinhold Company
135 West 50th Street
New York, NY 10020

A very comprehensive reference pertaining to the engineering and construction of molds for all kinds of plastic molding applications, from transfer to injection, and for the handling of thermoplastics to thermosets.

S-R-D-S
Standard Rate & Data Service
5201 Old Orchard Road
Skokie, IL 60676

The 'bible' of the publishing industry, Served up in several volumes for different publishing categories. Lists advertising rates, mechanical specifications and demographic information about thousands of publications. An excellent, annually updated reference.

School Masters Science
745 State Circle
Ann Arbor, MI 48104

Offers a $1.00 catalog of scientific apparatus, workbooks and other materials for schools and experimenters.

Scientific American
415 Madison Avenue
New York, NY 10017

A seasoned, contemporary, and articulate monthly coverage of a broad range of scientific subjects. Excellent depth.

Scientific Systems
P.O. Box 716
Amherst, NH 03031

Offers a $1.00 catalog of scientific apparatus, kits and plans.

Small Business Administration (SBA)
1441 L Street, NW
Washington, DC 20416

Consult Section XII of the Appendix for current SBA literature. SBA also sponsors a non-profit organization of retired businesspersons called 'SCORE.' They offer consultation, without charge, for persons starting new ventures. The SBA also engages in numerous other activities directed toward encouraging new business start-ups, including special studies on funding sources and capital formation, and participates in many special funding activities.

Small Business Innovation Research Program (SBIR)
National Science Foundation
1800 G Street NW, Room 1200
Washington, DC 20550

Consult with your SBA office for current information about areas in which the SBIR program is seeking to fund invention and innovation projects.

Small Business Sourcebook—Robert J. Elster
Gale Research Company
Book Tower
Detroit, MI 48226

ISSN 0883-3397. An exhaustive 2-volume, 1800 page guide to sources of information for small businesses, including associations, federal & state government agencies, consultants, small business incubator centers, educational institutions and programs, suppliers, reference works, statistical sources, trade periodicals, data bases, information services, libraries and research centers. Excellent!

Target Marketing
North American Publishing Co.
401 North Broad Street
Philadelphia, PA 19108

A 'controlled circulation' magazine for mail order professionals. A good source of continuing update on direct marketing ideas, supplies and equipment.

The Care And Feeding Of Ideas—James L. Adams
Addison-Wesley Publishing Co., Inc.
ISBN 0-201-10160-2

A provocative examination of the creative process with exercises in creative thinking and many tips and checklists for improving your inventive capabilities.

The Dartnell Corporation
4660 Ravenswood Avenue
Chicago, IL 60640

This firm publishes a number of business handbooks and other literature. The following are well recommended: *Sales Management Handbook, Direct Mail Order Handbook* (Hodgson), *Marketing Managers Handbook, Public Relations Handbook, Office Administration, Big Paybacks from Small Budget Advertising.*

The New Products Handbook
Dow Jones-Irwin Publishing
Homewood, IL 60430

A particularly detailed text examining the factors essential to new product success, including extensive evaluation, screening, and promotion elements. An extremely professional marketing text.

Thomas Register
Thomas Publishing Company
One Penn Plaza
New York, NY 10001

An encyclopedic set of volumes cataloging sources of virtually every conceivable kind of product from fastening devices, to motors, to ceramics, wood and literally thousands more. An excellent reference to American industry.

Tooling & Production
Huebcore Communications Inc.
P.O. Box 1074
Skokie, IL 60076-8074

A 'controlled circulation' magazine covering major areas of manufacturing from CAD/CAM to advanced methods of production. A very contemporary and useful publication for those contemplating manufacturing of a mechanical product.

Toy & Game Inventors of America
5813 McCart Avenue
Fort Worth, TX 76133

An association of independent toy inventors.

Toy Manufacturers of America, Inc.
Rm. 700
200 Fifth Avenue
New York, NY 10010

Offers useful information and suggestions for toy inventors. (Specifically does not buy inventions).

Toys, Hobbies & Crafts
545 Fifth Avenue
New York, NY 10017

Comprehensive coverage of this field and a particularly good continuing source reference.

Venture Magazine
Venture Magazine, Inc.
P.O. Box 3206
Harlan, IA 51537

A very slick, thoughtful range of articles covering a broad base of entrepreneurial interests.

VII. ADDITIONAL READING MATERIALS

The following are magazines, periodicals and advertising card decks which may be useful in researching marketplaces. Look for them in your library or write directly to the publisher. Most publishers will not send free copies, but generally will respond if you accompany your request for a sample copy with a check or money order for $3.00 to cover their mail and handling costs. Many magazines are "controlled circulation" which means that you must complete a qualification form. Controlled circulation magazines are generally free to qualified individuals whom that magazine's subscribers perceive as prospects for their goods or services.

Advertising Age
220 E. 42nd Street
New York, NY 10017

Advertising Specialty Inst.
1120 Wheeler Way
Langhorne, PA 19047

Advertising Techniques
10 E. 39th Street
New York, NY 10016

Agency Sales Magazine
MANA
P.O. Box 3467
Laguna Hills, CA 92654

Agri Marketing
5520 W. Touhy Ave.
Suite G
Skokie, IL 60077

Agribusiness News
P.O. Box 329
Titonka, IA 50480

American Industry
21 Russell Woods
Great Neck, NY 11021

American Journal of Science
217 Kline Geology Lab—Yale
P.O. Box 6666
New Haven, CT 06511

American Machinist
1221 Avenue Of The Americas
36th Floor
New York, NY 10020

Assembly Engineering
Hitchcock Publishing Co.
25 W. 550 Geneva Road
Wheaton, IL 60188

Assembly Engineering Postcard Pak
Hitchcock Publishing Co.
25 W. 550 Geneva Road
Wheaton, IL 60188

Automotive Age
6931 Van Nuys Blvd.
P.O. Box 2006
Van Nuys, CA 91405

Automotive Engineering
400 Commonwealth Drive
Warrendale, PA 15096

Aviation Mechanics Journal
1000 College View Drive
P.O. Box 36
Riverton, WY 82501

Biomedical Products Action Cards
Gordon Publications, Inc.
P.O. Box 1952
Dover, NJ 07801

Business & Industry
1200 35th Street
Suite 300
West Des Moines, IA 50265

Business Ideas
0151 Bloomfield Avenue
Clifton, NJ 07012

Business Marketing
220 E. 42nd Street
New York, NY 10017

Business Opportunities Journal
1021 Rosecrans Street
San Diego, CA 92106

Business Week
1221 Avenue of The Americas
New York, NY 10020

Byte
70 Main Street
Peterborough, NH 03458

Car Craft
8490 Sunset Blvd.
Los Angeles, CA 90069

Chemical Engineering
1221 Avenue Of The Americas
New York, NY 10020

Children's Business
7 East 12th Street
New York, NY 10003

Circuits Manufacturing
1050 Commonwealth Avenue
Boston, MA 02215

Creative
37 West 39th Street
New York, NY 10018

Creative Products News Quick Order Card
Cottage Communications
P.O. Box 584
Lake Forest, IL 60045

Design Engineering Postcard Program
Gordon Publications, Inc.
13 Emory Avenue
Randolph, NJ 07869

Design News Information Card Packs
Cahners Publishing Co.
275 Washington St.
Newton, MA 02158

Designfax
Huebner Publications, Inc.
6521 Davis Industrial Parkway
Solon, OH 44139

Designfax-Fax-Pax Cards
Huebner Publications, Inc.
6521 Davis Industrial Parkway
Solon, OH 44139

Direct Information Service Cards
International Scientific Commun.
P.O. Box 827
Fairfield, CT 06430

Direct Marketing Magazine
224 Seventh Street
Garden City, NY 11530

Doll Crafter/Tole World
Daisy Publishing
P.O. Box 67A
Mukilteo, WA 98275

Donald Moger Direct Marketing
P.O. Box 69219
750 North Kings Road
Los Angeles, CA 90069

Electrical Systems Design
One River Road
Cos Cob, CT 06807

Electronic Business
275 Washington Street
Newton, MA 02158

Electronic News
7 East 12th Street
New York, NY 10003

Electronic Products
645 Stewart Avenue
Garden City, NJ 11530

Electronics
1221 Avenue of The Americas
New York, NY 10020

Engineer's Digest
Walker-Davis Publications, Inc.
2500 Office Center
Willow Grove, PA 19090

Engineer's Digest Cards
Walker-Davis Publications, Inc.
2500 Office Center
Willow Grove, PA 19090

Evaluation Engineering
2504 N. Tamiami Trail
Nokomis, FL 33555

Execu-Deck/Industry Leaders Action Pac
Chilton Direct Marketing & List Mgmt
1 Chilton Way
Radnor, PA 19089

Experimental Techniques
Seven School Street
Bethel, CT 06801

Farm Equipment
1233 Janesville Avenue
Fort Atkinson, WI 53538

Financial Executives Action Cards
John Wiley & Sons, Inc.
605 Third Avenue
New York, NY 10158

Food & Drug Packaging
7500 Old Oak Blvd.
Cleveland, OH 44130

Food Eng. Prod. Twenty Postcard
Chilton Company
Chilton Way
Radnor, PA 19089

Food Technology
221 N. LaSalle Street
Chicago, IL 60601

Free Enterprise
824 E. Baltimore Street
Baltimore, MD 21202

Furniture Design & Mfg. Ad Cards
400 N. Michigan Avenue
Chicago, Il 60611

Game Trade News
1010 Vermont Avenue NW
Suite 910
Washington, DC 20005

Games Magazine
515 Madison Avenue
New York, NY 10022

Giftware News
Box 243
Deptford, NJ 08096

Hardware Age
Chilton Way
Radnor, PA 19089

Harvard Business Review
Teele Hall, Soldiers Field
Boston, MA 02163

High Tech Marketing
1460 Post Road E.
Westport, CT 06880

High Tech Times
Marketing Bulletin Board
3050 Calle Noguera
Santa Barber, CA 93105

High Technology
38 Commercial Wharf
Boston, MA 02110

Investment Action Cards
Direct Media, Inc.
220 Grace Church Street
Port Chester, NY 10573

Journal of Accountancy
1211 Avenue of The Americas
New York, NY 10026

Juvenile Merchandising
370 Lexington Avenue
New York, NY 10017

Licensing Book/Toy Book
264 W. 40th Street
New York, NY 10018

Machine Design-Pack Postcards
Penton/IPC, Penton Plaza
1111 Chester Avenue
Cleveland, OH 44114

Management Action Cards
Penton/IPC, Penton Plaza
1111 Chester Avenue
Cleveland, OH 44114

Management Technology
12 W. 21st Street
New York, NY 10011

Manager's Answer Card
Abarth Associates, Inc.
307 Asbury Road
Farmindale, NJ 07727

Manufacturing Engineering
Society of Manufacturing
Engineers
P.O. Box 930
Dearborn, MI 48121

Manufacturing Engineering Postcard
Society of Manufacturing
Engineers
P.O. Box 930
Dearborn, MI 48121

Marketing News
250 S. Wacker Drive, #200
Chicago, IL 60606

Materials Engineering
1111 Chester Avenue
Cleveland, OH 44114

Mechanical Engineering
345 E. 47th Street
New York, NY 10017

Metal Fabricating News
710 S. Main Street
Rockford, IL 61101

Metal Stamping
27027 Chardon Road
Richmond Heights, OH 44143

Metalworking Digest Action Cards
Gordon Publications, Inc.
13 Emery Avenue
Randolph, NJ 07869

Minority Business Entrepreneur
924 N. Market
Inglewood, CA 90302

Modern Applications News
2504 N. Tamiami Trail
Nokomis, FL 33555

Modern Electronics
76 N. Broadway
Hicksville, NY 11801

Modern Machine Shop Postcards
Gardner Publications, Inc.
6600 Clough Pike
Cincinnati, OH 45244

Modern Plastics
McGraw-Hill Publications
1221 Avenue of The Americas
New York, NY 10020

Modern Plastics Postcard Inquiry Service
McGraw-Hill Publications
1221 Avenue of the Americas
New York, NY 10020

Money Maker
5705 N. Lincoln Avenue
Chicago, IL 60659

Motor Trend
8490 Sunset Blvd.
Los Angeles, CA 90069

New Business Report
919 Third Avenue
New York, NY 10022

New Equipment Digest
1100 Superior Avenue
Cleveland, OH 54114

Omni
1965 Broadway
New York, NY 10023

Opportunity
Six North Michigan Avenue
Chicago, IL 60602

Packaging Digest PD Quick Pack
Delta Communications, Inc.
400 N. Michigan Ave.
Chicago, IL 60611

Packaging Information Cards
Cahners Publishing Co.
P.O. Box 5080
Des Plaines, IL 60018

Personal Eng. & Instrument.
Back Bay Annex
Box 903
Boston, MA 02117

Personal Robotics Magazine
P.O. Box 61047
Palo Alto, CA 94306

Plant Engineering
Technical Publishing
P.O. Box 1030
Barrington, IL 60010

Plant Engineering Postcard
Technical Publishing
P.O. Box 1030
Barrington, IL 60010

Plastics Design Forum
1129 E. 17th Avenue
Denver, CO 80218

Plastics Engineering
14 Fairfield Drive
Brookfield Center, CT 06805

Plastics Technology
633 Third Avenue
New York, NY 10017

Plastics World
275 Washington Street
Newton, MA 02158

Plastics World Information Cards
Cahners Publishing Co.
275 Washington St.
Newton, MA 02158

Playthings
51 Madison Avenue
New York, NY 10010

Prepared Foods Ad Card Paks
Gorman Publishing Co.
8750 W. Bryn Mawr Avenue
Chicago, IL 60631

Product Design & Development
Chilton Corporation
Chilton Way
Radnor, PA 19089

Product Design & Development Postpack
Chilton Corporation
Chilton Way
Radnor, PA 19089

Production Magazine
Box 557
5123 West Chester Pike
Edgemont, PA 19028

Research & Development
Technical Publishing
P.O. Box 1030
Barrington, IL 60010

Research & Development Ad Cards
Technical Publishing
P.O. Box 1030
Barrington, IL 60010

Research Management
100 Park Avenue, #3600
New York, NY 10017

RIA's Executive Decision Deck
Research Institute of America
589 5th Avenue
New York, NY 10016

Robotics Age
174 Concord Street
Peterborough, NH 03458

Robotics Today
One S M E Drive
P.O. Box 930
Dearborn, MI 48121

Robotics World
6255 Barfield Road
Atlanta, GA 30328

Robotics World Instant Action Postcard
Communication Channels, Inc.
18601 LBJ Freeway, Suite 240
Mesquite, TX 75150

Rubber & Plastics News Direct Response Card
Crain Automotive Group
34 North Hawkins Ave.
Akron, OH 44313

Sales & Marketing Ad Cards
Hughes Communications, Inc.
P.O. Box 197
Rockford, IL 61105

Sales & Marketing Management
633 Third Avenue
New York, NY 10017

Sales Motivators, Inc.
148 Third Street
P.O. Box 444
Baraboo, WI 53913

Science
1333 H. Street NW
Washington, DC 20005

Science & Industry Marketplace Card Mailer
Nature Publishing Co.
65 Bleeker Street
New York, NY 10012

Science Digest
888 Seventh Avenue
New York, NY 10106

Science News
1719 N Street N.W.
Washington, DC 20036

Science World
730 Broadway
New York, NY 10003

Self-Reliant
Route 72 & Jennings Road
Manahawkin, NJ 08050

Selling Direct
6255 Barfield Road
Atlanta, GA 30328

Small World
393 Seventh Avenue
New York, NY 10001

Solid State Technology
14 Vanderventer Avenue
Port Washington, NY 11050

Success
342 Madison Avenue,
21st Floor
New York, NY 10173

Technical Database Corp.
Computerized Manufacturing
Pac
P.O. Box 720
Conroe, TX 77305

Technology Review
Room 10-140
M.I.T.
Cambridge, MA 02139

*Telecommunications Products +
Technology Cards*
Penwell Publishing Co.
119 Russell Street
Littleton, MA 01460

The Best Report
350 Fifth Avenue, #4210
New York, NY 10118

The Business Owner
383 S. Broadway
Hicksville, NY 11801

The Robb Report
One Acton Place
Acton, MA 01720

The Sciences
2 East 63rd Street
New York, NY 10021

Tooling & Production
6521 Davis Industrial Pkwy
Solon, OH 44139

*Tooling & Production/Metlfax T &
Pfax*
Huebner Publications, Inc.
6521 Davis Industrial Parkway
Solon, OH 44139

*Tools of the Office Product Action
Card*
Dalton Communications, Inc.
1123 Broadway
New York, NY 60611

Toy & Hobby World
11 West 19th Street
New York, NY 10011

Toy Trade News
545 Fifth Avenue
New York, NY 10017

Ward's Auto World
28 W. Adams
Detroit, MI 48226

*What's New: For Science
Researchers*
Co-op Mailings
3004 Glenview Road
Wilmette, IL 60091

Wood Products Postcard Packets
Vance Publishing Corp.
P.O. Box 400
Prairie View, IL 60069

NOTE: Check the Yellow Pages of your local telephone directory or the nearest metropolitan directory for listings under 'Associations,' 'Marketing Consultants,' 'Business Consultants,' 'Patent Attorneys,' 'Tool Makers' and other categories of interest. Also, most regions of the country have business publications. Check this source as well as your local Chamber of Commerce.

VIII. LIBRARIES WITH PATENT RESOURCES

Excellent resources exist for researching patents, trademarks and copyrights. Some of these libraries are experimenting with a new computer system called CASSIS, which provides a direct link with the patent and trademark office. This enables researchers to access the federal computer and make quick searches of prior art.

In addition to the standard volumes of patents and trademarks, the reader will find the *Official Gazette, Rules of Patent Office Procedure*, microfilm and copying resources, and many other texts and materials covering the whole field of intellectual property. Personnel may change but names were current at publication.

Ms. Alta Beach
Science/Technology Reference Service
The New York State Library
Empire State Plaza
Albany, NY 12230

Mr. Eric Esau
Physical Sciences Library
Graduate Research Center
University of Massachusetts
Amherst, MA 01003

Ms. Margaret Bean
Engineering Transportation
Library – 312 UGL
The University of Michigan
Ann Arbor, MI 48109-1185

Mr. John F. Vandermolen
Science & Technology Dept.
Auburn University Libraries
Auburn University, AL 36849

Ms. Laurel Blewett
Bus. Admin./Govt. Documents
Troy H. Middleton Library
Louisiana State University
Baton Rouge, LA 70803

Ms. Marilyn McLean
Boston Public Library
Copley Square
Boston, MA 02117

Ms. Elizabeth Morrissett
Library/Montana College of
Mineral Science and Technology
Butte, MT 59701

Mr. Robert Poyer
Medical Univ. of South Carolina
171 Ashley Avenue
Charleston, SC 29425

Ms. Rosemary Dahmann
Science & Technology Dept.
Public Library of Cincinnati
800 Vine Street
Cincinnati, OH 45202-2071

Ms. Judy Erickson
Reference Services
Engineering & Physical Science Library
University of Maryland
College Park, MD 20742

Mr. Lawrence J. Perk
Special Materials Department
Ohio State University Libraries
1858 Neil Avenue Mall
Columbus, OH 43210

Mr. Robert Jackson
Business, Science, & Technology
Denver Public Library
1357 Broadway
Denver, CO 80203

Mr. Frank Adamovich
Patent Collection
The University Library
University Of New Hampshire
Durham, NH 03824

Ms. Barbara Kile
Div. of Govt. Publications
The Fondren Library/Rice Univ.
Houston, TX 77251-1892

Mr. Bruce B. Cox
Linda Hall Library
5109 Cherry Street
Kansas City, KS 64110

Ms. Mary Honeycutt
State Library Services
Arkansas State Library
One Capitol Mall
Little Rock, AR 72201-1081

Ms. Margaret Hayes
Tech. Rpts. & Patents Librarian
University of Wisconsin
215 North Randall Avenue
Madison, WI 53706

Mr. Edward Oswald
Business and Science Dept.
Miami-Dade Public Library
101 West Flagler Street
Miami FL 33130-2585

Ms. Edythe Abrahamson
Technology & Science Dept.
Minneapolis Public Library
300 Nicollet Mall
Minneapolis, MN 55401

Mr. Paul H. Murphy
Vanderbilt University
Science Library
419-21st Avenue South
Nashville, TN 37240-0007

Mr. Nicholas Patton
Science & Technology Div.
Newark Public Library
PO Box 630
Newark, NJ 07101

Mr. Charles Wilt
The Franklin Institute Library
20th Street & Parkway
Philadelphia, PA 19103

Mrs. Cheryl Hunt
Bus./Ind./Science/Patent Dept.
Providence Public Library
150 Empire Street
Providence, RI 02903

Mr. Steven D. Zink
Govt. Publ. Dept.
University Library
University of Nevada-Reno
Reno, NV 89557-0044

Mr. Thomas K. Anderson
California State Library
Library-Courts Building
PO Box 2037
Sacramento, CA 95809

Mr. Craig Smith
Reference & Circulation
Documents Section
Oregon State Library
Salem, OR 97310

Ms. Joanne Anderson
Science & Industry Section
San Diego Public Library
820 E. Street
San Diego, CA 92101

Ms. Jeanne Oliver
Reference Department
Illinois State Library
Centennial Building
Springfield, IL 62756

Ms. Mary-Jo DiMuccio
Patent Information
Clearinghouse
1500 Partridge Avenue
Building No. 7
Sunnyvale, CA 94087

Mrs. Mary Hubbard
Science/Technology Dept.
Toledo/Lucas City Library
325 Michigan Street
Toledo, OH 43624

Ms. Eulalie Brown
Govt. Publications & Maps Dept.
General Library
The University of New Mexico
Albuquerque, NM 87131

Ms. Colleen Sue McDonald
Anchorage Municipal Libraries
Z.J. Loussac Public Library
524 West 6th
Anchorage, AK 99501

Ms. Jean Kirkland
Dept. of Microforms
Prince Gilbert Memorial Library
Georgia Institute of Technology
Atlanta, GA 30332

Ms. Susan Ardis
McKinney Engineering Library
Room 1.3 ECJ
The University of Texas-Austin
Austin, TX 78712

Mrs. Rebecca Scarborough
Govt. Documents Dept.
Birmingham Public Library
2100 Park Place
Birmingham, AL 35203

Ms. Dorothy Solomon
Science & Technology Dept.
Buffalo & Erie City Public
Library
Lafayette Square
Buffalo, NY 14203

Ms. Karen Spak
Alliance College Library
Cambridge Springs, PA 16403

Ms. Diane A. Richmond
Science & Technology Info. Ctr.
Chicago Public Library
400 N. Michigan Avenue
Chicago, IL 60611

Mr. Siegfried Weinhold
Documents Collection
Cleveland Public Library
325 Superior Avenue
Cleveland, OH 44114-1271

Mr. W. David Gay
Evans Library
Documents Division
Texas A&M University
College Station, TX 77843-5000

Ms. Johanna Johnson
Dallas Public Library
1515 Young Street
Dallas, TX 75201

Ms. Barbara Klont
Technology Science Dept.
Detroit Public Library
5201 Woodward Avenue
Detroit, MI 48202

Ms. Christine Kitchens
Govt. Documents Dept.
Broward County Main Library
100 S. Andrews Avenue
Fort Lauderdale, FL 33301

Mr. Mark Leggett
Business, Science, & Tech. Div.
Indianapolis-Marion City Library
PO Box 211
Indianapolis, IN 46206

Mr. Alan Gould
Engineering Library
Nebraska Hall, 2nd Floor West
University of Nebraska-Lincoln
Lincoln, NE 68588-0410

Ms. Billie Connor
Science and Technology
Los Angeles Public Library
630 West Fifth Street
Los Angeles, CA 90071-2097

Ms. Barbara C. Shultz
Business/Science Department
Memphis & Shelby County
Library
1850 Peabody Avenue
Memphis, TN 38104

Mr. Theodore R. Cebula
Science & Business Coordinator
Milwaukee Public Library
814 West Wisconsin Avenue
Milwaukee, WI 53233

Ms. Donna M. Hanson
Science Librarian
University of Idaho Library
Moscow, ID 83843

Ms. Amanda Putnam
Patent Depository
Reference Department
University of Delaware Library
Newark, DE 19717-5267

Mr. Richard Hill
NYPL Annex
521 W. 43rd St.
New York, NY 10036-4396

Ms. Catherine M. Brosky
Science & Technology Dept.
Carnegie Library of Pittsburgh
4400 Forbes Avenue
Pittsburgh, PA 15213

Ms. Jean M. Porter
D.H. Hill Library
North Carolina State University
Box 7111
Raleigh, NC 27695-7111

Ms. Claire E. Hoffman
University Library Services
Virginia Commonwealth
University
901 Park Avenue
Richmond, VA 23284-0001

Ms. Stephanie Wunsh
Applied Science Department
St. Louis Public Library
1301 Olive Street
St. Louis, MO 63103

Ms. Julianne Hinz
Documents Division
Marriott Library
University of Utah
Salt Lake City, UT 84112

Ms. Vicki W. Phillips
Head Document Librarian
Oklahoma State Univ. Library
Stillwater, OK 74078

Ms. Diane H. Smith
Documents Section
C207 Pattee Library
Penn State University Libraries
University Park, PA 16802

IX. STATE ECONOMIC DEVELOPMENT OFFICES

Division of Business
Development
Department of Commerce &
Econ. Dev.
P.O. Box D
Juneau, AK 99811
(907) 465-2017

Industrial Financing Division
135 South Union St.
Suite 256
Montgomery, AL 36130
(205) 264-5441

Arkansas Industrial
Development Com.
One Capitol Mall, AIDC
Little Rock, AR 72201
(501) 371-1121

Development Finance Division
Arizona Department of
Commerce
1700 West Washington St.
Phoenix, AZ 85007
(602) 255-5705

Office of Small Business
Department of Commerce
1121 L. St., Suite 600
Sacramento, CA 95814
(916) 445-6545

Division of Commerce &
Development
1313 Sherman St.
Room 523
Denver, CO 80203
(303) 866-2205

Office of Small Business Services
Connecticut Dept. of Econ. Dev.
210 Washington St.
Hartford, CT 06106
(203) 566-4051

Office of Business and Economic
Development
1111 E. Street, N.W., Suite 700
Washington, DC 20004
(202) 727-6600

Delaware Development Office
99 King Highway
P.O. Box 1401
Dover, DE 19903
(302) 736-4271

Bureau of Business Assistance
Florida Department of
Commerce
G-26 Collins Building
Tallahassee, FL 32301
(904) 487-1314

State Office, Small Business Dev.
Center
Chicopee Bldg.
University of Georgia
Athens, GA 30602
(404) 542-1721

Small Business Information
Service
Department of Planning & Econ.
Dev.
P.O. Box 2359
Honolulu, HI 96804
(808) 548-7645

Small Business Office
Kansas Department of
Commerce
400 West 8th Street, Suite 500
Des Moines, IA 50309
(515) 281-3635

Iowa Department of Economic
Development
Department of Business Grants
& Loans
200 East Grand Avenue
Des Moines, IA 50309
(515) 281-3635

Department of Commerce
Economic Development & Data
Statehouse, Room 108
Boise, ID 83720
(208) 334-4719

Small Business Assistance
Bureau
Dept. of Commerce &
Community Affairs
620 East Adams, 5th Floor
Springfield, IL 62701
(217) 782-7500

Corporation for Innovation
Development
One North Capitol Ave., Suite
520
Indianapolis, IN 46204
(317) 635-7325

Small Business Office
Kansas Dept. of Commerce
400 West 8th Street, Suite 500
Topeka, KS 66603
(913) 296-3345

Kentucky Dev. Finance Auth.
2400 Capitol Plaza Tower
Frankfort, KY 40601
(502) 564-4554

Louisiana Small Business Equity
Corp.
4521 Jamestown Ave.—Suite 9
Baton Rouge, LA 70808
(504) 342-9213

Small Business Assistance
Division
Massachusetts Dept. of
Commerce
100 Cambridge St., 13th Floor
Boston, MA 02202
(617) 727-4005

Maryland Small Bus. Dev. Fin.
Auth.
World Trade Center
401 East Pratt St.
Baltimore, MD 21202
(301) 659-4270

Finance Authority of Maine
83 Western Ave.
P.O. Box 949
Augusta, ME 04330
(207) 623-3263

Michigan Strategic Fund
Michigan Department of
Commerce
P.O. Box 30234
Lansing, MI 48909
(517) 373-7550

Minnesota Small Business
Assistance Office
900 American Center Building
150 East Kellogg Boulevard
St. Paul, MN 55101
(612) 296-3871

Department of Economic
Development
Box 118
Jefferson City, MO 65102
(314) 751-4962

Mississippi Small Business
Clearinghouse
3825 Ridgewood Rd.
Jackson, MS 39211
(800) 521-7258

Business Assistance Division
Department of Commerce
1424 Ninth Avenue
Helena, MT 59620
(406) 444-3923

Small Business Development
Division
North Carolina Department of
Commerce
430 North Salisbury Street, Rm.
2019
Raleigh, NC 27611
(919) 733-7980

North Dakota Development
Commission
Liberty Memorial Building
Bismarck, ND 58501
(701) 224-2810

Nebraska Department of
Economic Development
Box 94666
3101 Centennial Mall South
Lincoln, NE 68509
(402) 471-3742

Office of Industrial Development
Department of Resources &
Econ. Dev.
P.O. Box 856
Concord, NH 03301
(603) 271-2591

Office of Small Business
Assistance
Dept. of Commerce & Econ.
Dev.
One West State Street, CN 823
Trenton, NJ 08625
(609) 984-4442

Economic Development &
Tourism Dept.
Joseph M. Montoya Building
1100 St. Francis Drive
Santa Fe, NM 87503
(505) 827-0305

Nevada Office of Community
Services
Small Business Revitalization
Program
1100 E. William, Suite 117
Carson City, NV 89710
(800) 992-0900 Ext. 4420

The Division for Small Business
New York State Dept. of
Commerce
230 Park Avenue
New York, NY 10169
(212) 309-0460

Small and Developing Business
Ohio Department of
Development
P.O. Box 1001
Columbus, OH 43266
(614) 466-4945

Established Industries Program
Small Business Representative
6601 North Broadway Ext., Suite
200
Oklahoma City, OK 73116
(409) 521-2182

Financial Programs
Oregon Economic Development
Dept.
595 Cottage St., NE
Salem, OR 97310
(503) 373-1215

Office of Small Business
Pennsylvania Dept. of Commerce
435 Forum Building
Harrisburg, PA 17120
(717) 787-2565

Department of Commerce
P.O. Box 4275
San Juan, PR 00905
(809) 724-0542

Small Business Development
Division
Rhode Island Dept. of Econ.
Dev.
7 Jackson Walkway
Providence, RI 02903
(401) 277-2601

Industry, Business and
Community Serv.
South Carolina State Dev. Board
P.O. Box 927
Columbia, SC 29202
(803) 734-1400

Governor's Office of Economic
Development
Capitol Lake Plaza
711 Wells
Pierre, SD 57501
(605) 773-5032

Small Business Office
Dept. of Econ. & Commun. Dev.,
7th Floor
Rachel Jackson Building
Nashville, TN 37219
(800) 872-7201

Texas Economic Development
Commission
P.O. Box 12728
Austin, TX 78711
(512) 472-5059

Division of Business & Econ.
Development
6136 State Office Building
Salt Lake City, UT 84114
(801) 533-5325

Office of Small Business and Fin.
Serv.
Virginia Department of
Economic Dev.
1000 Washington Building
Richmond, VA 23219
(804) 786-3791

Small Business Development
Agency
P.O. Box 2058
St. Thomas, VI 00801
(809) 774-8784

Economic Development
Department
109 State St.
Montpelier, VT 05602
(802) 828-3221

Development Services Division
Department of Trade & Econ.
Dev.
101 General Administration
Bldg., AX-13
Olympia, WA 98504
(206) 753-5630

Wisconsin Department of
Development
P.O. Box 2970
Madison, WI 53707
(608) 266-0562

Small Business Development
Center Div.
Governor's Office of Com. &
Ind. Dev.
State Capitol
Charleston, WV 25305
(800) 225-5982

Wyoming Development
Corporation
P.O. Box 3599
Casper, WY 82602
(307) 234-5351

X. DIRECTORY OF SBICS AND SEED CAPITAL NETWORKS

ACTIVE SBICS

Alabama

First SBIC of Alabama
David Delaney, President
16 Midtown Park East
Mobile, AL 36606
(205) 476-0700

Hickory Venture Capital Corp.
J. Thomas Noojin, President
699 Gallatin Street, Suite A-2
Huntsville, AL 35801
(205) 539-1931

Remington Fund, Inc. (The)
Lana Sellers, President
1927 First Avenue North
Birmingham, AL 35202
(205) 324-7709

Alaska

Alaska Bus. Investment Corp.
James Cloud, Vice President
301 West Northern Lights Blvd.
Mail: P.O. Box 100600;
Anchorage, AK 99510
(907) 278-2071

Arizona

Northwest Venture Partners
(Main Office: Minneapolis, MN)
88777 E. Via de Ventura
Suite 335
Scottsdale, AZ 85258
(602) 483-8940

Norwest Growth Fund, Inc.
(Main Office: Minneapolis, MN)
88777 E. Via de Ventura
Suite 335
Scottsdale, AZ 85258
(602) 483-8940

Rocky Mountain Equity Corp.
Anthony J. Nicoli, President
4530 Central Avenue
Phoenix, AZ 85012
(602) 274-7534

Valley National Investors, Inc.
John M. Holliman III, V.P. &
Manager
201 North Central Avenue, Suite
900
Phoenix, AZ 85004
(602) 261-1577

Wilbur Venture Capital Corp.
Jerry F. Wilbur, President
4575 South Palo Verde, Suite
305
Tucson, AZ 85714
(602) 747-5999

Arkansas

Small Business Inv. Capital, Inc.
Charles E. Toland, President
10003 New Benton Hwy.
Mail: P.O. Box 3627
Little Rock, AR 72203
(501) 455-6599

Southern Ventures, Inc.
Jeffrey A. Doose, President &
Director
605 Main Street, Suite 202
Arkadelphia, AR 71923
(501) 246-9627

California

AMF Financial, Inc.
William Temple, Vice President
4330 La Jolla Village Drive
Suite 110
San Diego, CA 92122
(619) 546-0167

Atalanta Investment Company,
Inc.
(Main Office: New York, NY)
141 El Camino Drive
Beverly Hills, CA 90212
(213) 273-1730

BNP Venture Capital
Corporation
Edgerton Scott II, President
3000 Sand Hill Road
Building 1, Suite 125
Menlo Park, CA 94025
(415) 854-1084

Bancorp Venture Capital, Inc.
Arthur H. Bernstein, President
11812 San Vicente Boulevard
Los Angeles, CA 90049
(213) 820-7222

BankAmerica Ventures, Inc.
Groups 26 & 38
Patrick Topolski, President
555 California Street
San Francisco, CA 94104
(415) 953-3001

CFB Venture Capital
Corporation
Richard J. Roncaglia, Vice
President
530 B. Street, Third Floor
San Diego, CA 92101
(619) 230-3304

CFB Venture Capital
Corporation
(Main Office: San Diego, CA)
350 California Street, Mezzanine
San Francisco, CA 94104
(415) 445-0594

Citicorp Venture Capital, Ltd.
(Main Office: New York, NY)
2 Embarcadero Place
2200 Geny Road, Suite 203
Palo Alto, CA 94303
(415) 424-8000

City Ventures, Inc.
Warner Heineman, Vice
Chairman
400 N. Roxbury Drive
Beverly Hills, CA 90210
(213) 550-5709

Crosspoint Investment Corp.
Max Simpson, Pres. & Chief F.O.
1951 Landings Drive
Mountain View, CA 94043
(415) 968-0930

Developers Equity Capital Corp.
Larry Sade, Chmn of the Board
1880 Century Park East
Suite 311
Los Angeles, CA 90067
(213) 277-0330

Draper Associates, a California
L.P.
Bill Edwards, President
c/o Timothy C. Draper
3000 Sand Hill Road, Bldg 4,
#235
Menlo Park, CA 94025
(415) 854-1712

First Interstate Capital, Inc.
Ronald J. Hall, Managing Dir.
5000 Birch Street, Suite 10100
Newport Beach, CA 92660
(714) 253-4360

First SBIC of California
Tim Hay, President
650 Town Center Drive
Seventeenth Floor
Costa Mesa, CA 92626
(714) 556-1964

First SBIC of California
(Main Office: Costa Mesa, CA)
5 Palo Alto Square, Suite 938
Palo Alto, CA 94306
(415) 424-8011

First SBIC of California
(Main Office: Costa Mesa, CA)
155 North Lake Avenue, Suite
1010
Pasadena, CA 91109
(818) 304-3451

G C & H Partners
James C. Gaither, General
Partner
One Maritime Plaza, 20th Floor
San Francisco, CA 94110
(415) 981-5252

Hamco Capital Corp.
William R. Hambrecht, President
235 Montgomery Street
San Francisco, CA 94104
(415) 576-3635

Imperial Ventures, Inc.
H. Wayne Snavely, President
9920 South Lacienega Blvd.
Mail: P.O. Box 92991; L.A. 90009
Inglewood, CA 90301
(213) 417-5888

Jupiter Partners
John M. Bryan, President
600 Montgomery Street
35th Floor
San Francisco, CA 94111
(415) 421-9990

Latigo Capital Partners, II
Robert A. Peterson, General
Partner
1800 Century Park East, Suite
430
Los Angeles, CA 90067
(213) 556-2666

Marwit Capital Corp.
Martin W. Witte, President
180 Newport Center Drive
Suite 200
Newport Beach, CA 92660
(714) 640-6234

Merrill Pickard Anderson & Eyre
Inc.
Steven L. Merrill, President
Two Palo Alto Square, Suite 425
Palo Alto, CA 94306
(415) 856-8880

Metropolitan Venture Company,
Inc.
Rudolph J. Lowy, Chairman of
the Board
5757 Wilshire Blvd.
Suite 670
Los Angeles, CA 90036
(213) 938-3488

New West Partners II
Timothy P. Haidinger, Manager
4350 Executive Drive, Suite 206
San Diego, CA 92121
(619) 457-0723

New West Partners II
(Main Office: San Diego, CA)
4600 Campus Drive, Suite 103
Newport Beach, CA 92660
(714) 756-8940

PBC Venture Capital Inc.
Henry L. Wheeler, Manager
1408−18th Street
Mail: P.O. Box 6008; Bakersfield
93386
Bakersfield, CA 93301
(805) 395-3555

Peerless Capital Company, Inc.
Robert W. Lautz, Jr., President
675 South Arroyo Parkway
Suite 320
Pasadena, CA 91105
(818) 577-9199

Ritter Partners
William C. Edwards, President
150 Isabella Avenue
Atherton, CA 94025
(415) 854-1555

Round Table Capital Corp.
Richard Dumke, President
655 Montgomery Street, Suite
700
San Francisco, CA 94111
(415) 392-7500

San Joaquin Capital Corporation
Chester Troudy, President
1415 18th Street, Suite 306
Mail: P.O. Box 2538
Bakersfield, CA 93301
(805) 323-7581

Seaport Ventures, Inc.
Michael Stopler, President
525 B Street, Suite 630
San Diego, CA 92101
(619) 232-4069

Union Venture Corp.
Jeffrey Watts, President
445 South Figueroa Street
Los Angeles, CA 90071
(213) 236-4092

VK Capital Company
Franklin Van Kasper, G.P.
50 California Street, Suite 2350
San Francisco, CA 94111
(415) 391-5600

Vista Capital Corp.
Frederick J. Howden, Jr.,
Chairman
5080 Shoreham Place, Suite 202
San Diego, CA 92122
(619) 453-0780

Walden Capital Partners
Arthur S. Berliner, President
750 Battery Street, Seventh Floor
San Francisco, CA 94111
(415) 391-7225

Wells Fargo Capital Corporation
Ms. Sandra J. Menichelli, V.P. &
G.M.
420 Montgomery Street, 9th
Floor
San Francisco, CA 94163
(415) 396-2059

Westamco Investment Company
Leonard G. Muskin, President
8929 Wilshire Blvd., Suite 400
Beverly Hills, CA 90211
(213) 652-8288

Colorado

Associated Capital Corporation
Rodney J. Love, President
4891 Independence Street, Suite
201
Wheat Ridge, CO 80033
(303) 420-8155

UBD Capital, Inc.
Allan R. Haworth, President
1700 Broadway
Denver, CO 80274
(303) 863-6329

Connecticut

AB SBIC, Inc.
Adam J. Bozzuto, President
275 School House Road
Cheshire, CT 06410
(203) 272-0203

All State Venture Capital Corp.
Ceasar N. Anquillare, President
The Bishop House
32 Elm Street, P.O. Box 1629
New Haven, CT 06506
(203) 787-5029

Capital Impact Corp.
William D. Starbuck, President
961 Main Street
Bridgeport, CT 06601
(203) 384-5670

Capital Resource Co. of
Connecticut
I. Martin Fierberg, Managing
Partner
699 Bloomfield Avenue
Bloomfield, CT 06002
(203) 243-1114

Dewey Investment Corp.
George E. Mrosek, President
101 Middle Turnpike West
Manchester, CT 06040
(203) 649-0654

First Connecticut SBIC
David Engelson, President
177 State Street
Bridgeport, CT 06604
(203) 366-4726

First New England Capital, L.P.
Richard C. Klaffky, President
255 Main Street
Hartford, CT 06106
(203) 728-5200

Marcon Capital Corp.
Martin A. Cohen, President
49 Riverside Avenue
Westport, CT 06880
(203) 226-6893

Northeastern Capital Corp.
Joseph V. Ciaburri, Chmn/CEO
209 Church Street
New Haven, CT 06510
(203) 865-4500

Regional Financial Ent., L.P.
Robert M. Williams, Managing
Partber
36 Grove Street
New Canaan, CT 06840
(203) 966-2800

SBIC of Connecticut Inc. (The)
Kenneth F. Zarrilli, President
1115 Main Street
Bridgeport, CT 06603
(203) 367-3282

Delaware

Morgan Investment Corporation
William E. Pike, Chairman
902 Market Street
Wilmington, DE 19801
(302) 651-2500

District of Columbia

Allied Investment Corporation
David J. Gladstone, President
1666 K Street, N.W., Suite 901
Washington, DC 20006
(202) 331-1112

American Security Capital Corp.,
Inc.
William G. Tull, President
730 Fifteenth Street, N.W.
Washington, DC 20013
(202) 624-4843

DC Bancorp Venture Capital Co.
Allan A. Weissburg, President
1801 K Street, N.W.
Washington, DC 20006
(202) 955-6970

Washington Ventures, Inc.
Kenneth A. Swain, President
1320 18th Street, N.W., Suite 300
Washington, DC 20036
(202) 895-2560

Florida

Allied Investment Corporation
(Main Office: Washington, DC)
Executive Office Cntr, Suite 305
2770 N. Indian River Blvd.
Vero Beach, FL 32960
(407) 778-5556

First North Florida SBIC
J.B. Higdon, President
1400 Gadsden Street
P.O. Box 1021
Quincy, FL 32351
(904) 875-2600

Gold Coast Capital Corporation
William I. Gold, President
3550 Biscayne Blvd., Room 601
Miami, FL 33137
(305) 576-2012

J & D Capital Corp.
Jack Carmel, President
12747 Biscayne Blvd.
North Miami, FL 33181
(305) 893-0303

Market Capital Corp.
E. E. Eads, President
1102 North 28th Street
P.O. Box 22667
Tampa, FL 33630
(813) 247-1357

Quantum Capital Partners, Ltd.
Michael E. Chaney, President
2400 East Commercial Blvd,
Suite 814
Fort Lauderdale, FL 33308
(305) 776-1133

Southeast Venture Capital
Limited Inc
James R. Fitzsimons, Jr., Pres.
3250 Miami Center
100 Chopin Plaza
Miami, FL 33131
(305) 379-2005

Western Financial Capital
Corporation
(Main Office: Dallas, TX)
1380 N.E. Miami Gardens Drive
Suite 225
N. Miami Beach, FL 33179
(305) 949-5900

Georgia

Investor's Equity, Inc.
I. Walter Fisher, President
2629 First National Bank Tower
Atlanta, GA 30383
(404) 523-3999

North Riverside Capital
Corporation
Tom Barry, President
50 Technology Park/Atlanta
Norcross, GA 30092
(404) 446-5556

Hawaii

Bancorp Hawaii SBIC
James D. Evans, Jr., President
111 South King Street
Suite 1060
Honolulu, HI 96813
(808) 521-6411

Illinois

ANB Venture Corporation
Kurt L. Liljedahl,
Exec. Vice-President
33 North LaSalle Street
Chicago, IL 60690
(312) 855-1554

Alpha Capital Venture Partners,
L.P.
Andrew H. Kalnow, General
Partner
Three First National Plaza, 14th
Floor
Chicago, IL 60602
(312) 372-1556

Business Ventures, Incorporated
Milton Lefton, President
20 North Wacker Drive, Suite
550
Chicago, IL 60606
(312) 346-1580

Continental Illinois Venture
Corp.
John L. Hines, President
209 South LaSalle Street
Mail: 231 South LaSalle Street
Chicago, IL 60693
(312) 828-8023

First Capital Corp. of Chicago
John A. Canning, Jr., President
Three First National Plaza
Suite 1330
Chicago, IL 60670
(312) 732-5400

Frontenac Capital Corporation
David A. R. Dullum, President
208 South LaSalle Street,
Room 1900
Chicago, IL 60604
(312) 368-0047

Heller Equity Capital
Corporation
Robert E. Koe, President
200 North LaSalle Street
10th Floor
Chicago, IL 60601
(312) 621-7200

Mesirow Capital Partners SBIC,
Ltd.
James C. Tyree, President of
C.G.P.
1355 LaSalle Street, Suite 3910
Chicago, IL 60603
(312) 443-5773

Walnut Capital Corp.
Burton W. Kanter, Chairman of
the Board
208 South LaSalle Street
Chicago, IL 60604
(312) 346-2033

Indiana

1st Source Capital Corporation
Eugene L. Cavanaugh, Jr., Vice
President
100 North Michigan Street
South Bend, IN 46601
Mail: P.O. Box 1602
South Bend, IN 46634
(219) 236-2180

Circle Ventures, Inc.
Robert Salyers, President
2502 Roosevelt Avenue
Indianapolis, IN 46218
(317) 636-7242

Equity Resource Company, Inc.
Michael J. Hammes, Vice
President
One Plaza Place
202 South Michigan Street
South Bend, IN 46601
(219) 237-5255

Raffensperger Hughes Venture
Corp.
Samuel B. Sutphin, President
20 North Meridian Street
Indianapolis, IN 46204
(317) 635-4551

White River Capital Corporation
Thomas D. Washburn, President
500 Washington Street
Mail: P.O. Box 929
Columbus, IN 47201
(812) 372-0111

Iowa

MorAmerica Capital
Corporation
David R. Schroder, Vice
President
800 American Building
Cedar Rapids, IA 52401
(319) 363-8249

Kansas

Kansas Venture Capital, Inc.
Larry J. High, President
First National Bank Tower,
Suite 825
One Townsite Plaza
Topeka, KS 66603
(913) 233-1368

Kentucky

Financial Opportunities, Inc.
Gary Duerr, Manager
6060 Dutchman's Lane
Mail: PO Box 35710; Louisville,
KY 40232
Louisville, KY 40205
(502) 451-3800

Mountain Ventures, Inc.
Jerry A. Rickett,
Exec. Vice President
London Bank & Trust Building
400 S. Main Street, Fourth Floor
London, KY 40741
(606) 864-5175

Wilbur Venture Capital Corp.
(Main Office: Tucson, AZ)
400 Fincastle Building
3rd & Broadway
Louisville, KY 40202
(502) 585-1214

Louisiana

Capital for Terrebonne, Inc.
Hartwell A. Lewis, President
27 Austin Drive
Houma, LA 70360
(504) 868-3930

Louisiana Equity Capital
Corporation
G. Lee Griffin, President
451 Florida Street
Baton Rouge, LA 70821
(504) 389-4421

Maine

Maine Capital Corp.
David M. Coit, President
Seventy Center Street
Portland, ME 04101
(207) 772-1001

Maryland

First Maryland Capital, Inc.
Joseph A. Kenary, President
107 West Jefferson Street
Rockville, MD 20850
(301) 251-6630

Greater Washington
Investments, Inc.
Don A. Christensen, President
5454 Wisconsin Avenue
Chevy Chase, MD 20815
(301) 656-0626

Jiffy Lube Capital Corporation
Eleanor C. Harding, President
6000 Metro Drive
Mail: PO Box 17223; Baltimore
21203-7223
Baltimore, MD 21215
(301) 764-3234

Massachusetts

Advent Atlantic Capital
Company, L.P.
David D. Croll, Managing
Partner
45 Milk Street
Boston, MA 02109
(617) 338-0800

Advent IV Capital Company
David D. Croll, Managing
Partner
45 Milk Street
Boston, MA 02109
(617) 338-0800

Advent Industrial Capital
Company, L.P.
David D. Croll, Managing
Partner
45 Milk Street
Boston, MA 02109
(617) 338-0800

Advent V Capital Company L.P.
David D. Croll, Managing
Partner
45 Milk Street
Boston, MA 02109
(617) 338-0800

Atlas II Capital Corporation
Joost E. Tjaden, President
101 Federal Street, 4th Floor
Boston, MA 02110
(617) 951-9420

BancBoston Ventures,
Incorporated
Paul F. Hogan, President
100 Federal Street
Mail: P.O. Box 2016 Stop 01-31-
08
Boston, MA 02110
(617) 434-2441

Bever Capital Corp.
Joost E. Tjaden, President
101 Federal Street, 4th Floor
Boston, MA 02110
(617) 951-9420

Boston Hambro Capital
Company
Edwin Goodman, President of
Corp. G.P.
160 State Street, 9th Floor
Boston, MA 02109
(617) 523-7767

Business Achievement
Corporation
Michael L. Katzeff, President
1172 Beacon Street, Suite 202
Newton, MA 02161
(617) 965-0550

Chestnut Capital International II
LP
David D. Croll, Managing
Partner
45 Milk Street
Boston, MA 02109
(617) 338-0800

Chestnut Street Partners, Inc.
David D. Croll, President
45 Milk Street
Boston, MA 02109
(617) 574-6763

First Capital Corp. of Chicago
(Main Office: Chicago, IL)
133 Federal Street, 6th Floor
Boston, MA 02110
(617) 542-9185

First United SBIC, Inc.
Alfred W. Ferrara, Vice Pres.
135 Will Drive
Canton, MA 02021
(617) 828-6150

Fleet Venture Resources, Inc.
(Main Office: Providence, RI)
Carlton V. Klein, Vice-President
60 State Street
Boston, MA 02109
(617) 367-6700

Mezzanine Capital Corporation
David D. Croll, President
45 Milk Street
Boston, MA 02109
(617) 574-6752

Milk Street Partners, Inc.
Richard H. Churchill, Jr.,
President
45 Milk Street
Boston, MA 02109
(617) 574-6723

Monarch-Narragansett Ventures,
Inc.
George W. Siguler, President
One Financial Plaza
Springfield, MA 01102
(413) 781-3000

New England Capital Corp.
Z. David Patterson, Vice Pres.
One Washington Mall, 7th Floor
Boston, MA 02108
(617) 573-6400

Northeast SBI Corp.
Joseph Mindick, Treasurer
16 Cumberland Street
Boston, MA 02115
(617) 267-3983

Orange Nassau Capital
Corporation
Joost E. Tjaden, President
101 Federal Street, 4th Floor
Boston, MA 02110
(617) 951-9420

Pioneer Ventures Limited
Partnership
Christopher W. Lynch, Managing
Partner
60 State Street
Boston, MA 02109
(617) 742-7825

Shawmut National Capital
Corporation
Steven James Lee, President
One Federal Street, 30th Floor
Boston, MA 02211
(617) 556-4700

Stevens Capital Corporation
Edward Capuano, President
168 Stevens Street
Fall River, MA 02721
(617) 679-0044

UST Capital Corp.
Walter Dick, President
40 Court Street
Boston, MA 02108
(617) 726-7137

Vadus Capital Corp.
Joost E. Tjaden, President
101 Federal Street, 4th Floor
Boston, MA 02110
(617) 951-9420

Michigan

Michigan Tech Capital Corp.
Clark L. Pellegrini, President
Technology Park
601 West Sharon Avenue
P.O. Box 364
Houghton, MI 49931
(906) 487-2970

Minnesota

FBS SBIC, Limited Partnership
John M. Murphy, Jr., Managing
Agent
1100 First Bank Place East
Minneapolis, MN 55480
(612) 370-4764

North Star Ventures II, Inc.
Terrence W. Glarner, President
150 South Fifth Street
Suite 3400
Minneapolis, MN 55402
(612) 333-1133

Northland Capital Venture
Partnership
George G. Barnum, Jr., President
613 Missabe Building
Duluth, MN 55802
(218) 722-0545

Northwest Venture Partners
Robert F. Zicarelli, Managing
G.P.
2800 Piper Jaffray Tower
222 South Ninth Street
Minneapolis, MN 55402
(612) 372-8770

Norwest Growth Fund, Inc.
Daniel J. Haggerty, President
2800 Piper Jaffray Tower
222 South Ninth Street
Minneapolis, MN 55402
(612) 372-8770

Shared Ventures, Inc.
Howard W. Weiner, President
6550 York Avenue, South
Suite 419
Edina, MN 55435
(612) 925-3411

Misssouri

Bankers Capital Corp.
Raymond E. Glasnapp, President
3100 Gillham Road
Kansas City, MO 64109
(816) 531-1600

Capital for Business, Inc.
James B. Hebenstreit, President
1000 Walnut, 18th Floor
Kansas City, MO 64106
(816) 234-2357

Capital for Business, Inc.
(Main Office: Kansas City, MO)
11 South Meramec, Suite 804
St. Louis, MO 63105
(314) 854-7427

MBI Venture Capital Investors,
Inc.
Anthony Sommers, President
850 Main Street
Kansas City, MO 64105
(816) 471-1700

MorAmerica Capital
Corporation
(Main Office: Cedar Rapids, IA)
911 Main Street, Suite 2724A
Commerce Tower Building
Kansas City, MO 64105
(816) 842-0114

United Missouri Capital
Corporation
Joe Kessinger, Manager
1010 Grand Avenue
Mail: P.O. Box 419226; K.C., MO
64141
Kansas City, MO 64106
(816) 556-7333

Nebraska

First of Nebraska Investment
Corp.
Dennis O'Neal, Managing Off.
One First National Center
Suite 701
Omaha, NE 68102
(402) 633-3585

United Financial Resources
Corp.
Dennis L. Schulte, Manager
6211 L Street
Mail: P.O. Box 1131
Omaha, NE 68101
(402) 734-1250

Nevada

Enterprise Finance Cap
Development Corp.
Robert S. Russell, Sr., President
First Interstate Bank of Nevada
Bldg.
One East First Street, Suite 1100
Reno, NV 89501
(702) 329-7797

New Hampshire

VenCap, Inc.
Richard J. Ash, President
1155 Elm Street
Manchester, NH 03101
(603) 644-6100

New Jersey

Bishop Capital, L.P.
Charles J. Irish
58 Park Place
Newark, NJ 07102
(201) 623-0171

ESLO Capital Corp.
Leo Katz, President
212 Wright Street
Newark, NJ 07114
(201) 242-4488

First Princeton Capital
Corporation
Michael D. Feinstein, President
Five Garret Mountain Plaza
West Paterson, NJ 07424
(201) 278-8111

Monmouth Capital Corp.
Eugene W. Landy, President
125 Wycoff Road
Midland National Bank Bldg.
P.O. Box 335
Eatontown, NJ 07724
(201) 542-4927

Tappan Zee Capital Corporation
Karl Kirschner, President
201 Lower Notch Road
Little Falls, NJ 07424
(201) 256-8280

Unicorn Ventures II, L.P.
Frank P. Diassi, General Partner
6 Commerce Drive
Cranford, NJ 07016
(201) 276-7880

Unicorn Ventures, Ltd.
Frank P. Diassi, President
6 Commerce Drive
Cranford, NJ 07016
(201) 276-7880

United Jersey Venture Capital,
Inc.
Stephen H. Paneyko, President
301 Carnegie Center
P.O. Box 2066
Princeton, NJ 08540
(609) 987-3490

New Mexico

Albuquerque SBIC
Albert T. Ussery, President
501 Tijeras Avenue, N.W.
P.O. Box 487
Albuquerque, NM 87103
(505) 247-0145

Equity Capital Corp.
Jerry A. Henson, President
119 East Marcy Street, Suite 101
Santa Fe, NM 87501
(505) 988-4273

Southwest Capital Investments,
Inc.
Martin J. Roe, President
The Southwest Building
3500-E Comanche Road, N.E.
Albuquerque, NM 87107
(505) 884-7161

United Mercantile Capital Corp.
Joe Justice, General Manager
2400 Louisiana Blvd., Bldg. 4,
Suite 101
Mail: P.O. Box 37487;
Albuquerque 87176
Albuquerque, NM 87110
(505) 883-8201

New York

767 Limited Partnership
H. Wertheim and H. Mallement,
G.P.
767 Third Avenue
New York, NY 10017
(212) 838-7776

ASEA-Harvest Partners II
Harvey Wertheim, General
Partner
767 Third Avenue
New York, NY 10017
(212) 838-7776

American Commercial Capital
Corporation
Gerald J. Grossman, President
310 Madison Avenue, Suite 1304
New York, NY 10017
(212) 986-3305

American Energy Investment
Corp.
John J. Hoey, Chmn of the Board
645 Fifth Avenue, Suite 1900
New York, NY 10022
(212) 688-7307

Amev Capital Corp.
Martin Orland, President
One World Trade Center
50th Floor
New York, NY 10048
(212) 775-9100

Atalanta Investment Company,
Inc.
L. Mark Newman, Chairman of
the Board
450 Park Avenue
New York, NY 10022
(212) 832-1104

BT Capital Corp.
James G. Hellmuth, Deputy
Chairman
280 Park Avenue, 10 West
New York, NY 10017
(212) 850-1916

Boston Hambro Capital
Company
(Main Office: Boston, MA)
17 East 71st Street
New York, NY 10021
(212) 288-9106

Bridger Capital Corp.
Seymour L. Wane, President
645 Madison Avenue, Suite 810
New York, NY 10022
(212) 888-4004

CMNY Capital, L.P.
Robert Davidoff, General
Partner
77 Water Street
New York, NY 10005
(212) 437-7078

Central New York SBIC (The)
Albert Wertheimer, President
351 South Warren Street
Syracuse, NY 13202
(315) 478-5026

Chase Manhattan Capital
Corporation
Gustav H. Koven, President
1 Chase Manhattan Plaza, 23rd
Floor
New York, NY 10081
(212) 552-6275

Chemical Venture Capital
Associates
Jeffrey C. Walker, Managing
Gen. Partner
277 Park Avenue, 10th Floor
New York, NY 10172
(212) 310-7578

Citicorp Venture Capital, Ltd.
William Comfort, Chairman of
the Board
399 Park Avenue, 6th Floor
New York, NY 10043
(212) 559-1127

Clinton Capital Corp.
Mark Scharfman, President
79 Madison Avenue, Suite 800
New York, NY 10016
(212) 696-4334

Croyden Capital Corp.
Lawrence D. Gorfinkle,
President
45 Rockefeller Plaza, Suite 2165
New York, NY 10111
(212) 974-0184

Diamond Capital Corp.
Steven B. Kravitz, President
805 Third Avenue, Suite 1100
New York, NY 10017
(212) 838-1255

Edwards Capital Company
Edward H. Teitlebaum, President
215 Lexington Avenue, Suite 805
New York, NY 10016
(212) 686-2568

F/N Capital Limited Partnership
Raymond A. Lancaster,
President
One Norstar Plaza
Albany, NY 12207
(518) 447-4050

Fairfield Equity Corp.
Matthew A. Berdon, President
200 East 42nd Street
New York, NY 10017
(212) 867-0150

Technology, Inc.
Sandford R. Simon, President &
Director
515 Madison Avenue
New York, NY 10022
(212) 688-9828

Fifty-Third Street Ventures, L.P.
Patricia Cloherty & Dan Tessler,
G.P.
155 Main Street
Cold Spring, NY 10516
(914) 265-5167

Franklin Corporation SBIC
(The)
Norman S. Strobel, President
767 Fifth Avenue
G.M. Building, 23rd Floor
New York, NY 10153
(212) 486-2323

Fundex Capital Corp.
Howard Sommer, President
525 Northern Blvd.
Great Neck, NY 11021
(516) 466-8551

GHW Capital Corp.
Philip Worlitzer, Vice President
25 West 45th Street, Suite 707
New York, NY 10036
(212) 869-4584

Genesee Funding, Inc.
A. Keene Bolton, President,
CEO
100 Corporate Woods
Rochester, NY 14623
(716) 272-2332

Hanover Capital Corp. (The)
Geoffrey T. Selzer, President
150 East 58th Street, Suite 2710
New York, NY 10155
(212) 980-9670

Intergroup Venture Capital
Corp.
Ben Hauben, President
230 Park Avenue
New York, NY 10017
(212) 661-5428

Interstate Capital Company, Inc.
David Scharf, President
380 Lexington Avenue
New York, NY 10017
(212) 986-7333

Irving Capital Corp.
Andrew McWethy, President
1290 Avenue of the Americas
New York, NY 10104
(212) 408-4800

Kwiat Capital Corp.
Sheldon F. Kwiat, President
576 Fifth Avenue
New York, NY 10036
(212) 391-2461

M & T Capital Corp.
William Randon, President
One M & T Plaza
Buffalo, NY 14240
(716) 842-5881

MH Capital Investors, Inc.
Edward L. Kock III, President
270 Park Avenue
New York, NY 10017
(212) 286-3222

Multi-Purpose Capital
Corporation
Eli B. Fine, President
5 West Main Street, Room 207
Elmsford, NY 10523
(914) 347-2733

NYBDC Capital Corp.
Robert W. Lazar, President
41 State Street
Albany, NY 12207
(518) 463-2268

NYSTRS/NV Capital, Limited
Partnership
Raymond A. Lancaster,
President
One Norstar Plaza
Albany, NY 12207
(518) 447-4050

NatWest USA Capital
Corporation
Orville G. Aarons, General
Manager
175 Water Street
New York, NY 10038
(212) 602-1200

Norstar Capital, Inc.
Raymond A. Lancaster,
President
One Norstar Plaza
Albany, NY 12207
(518) 447-4043

Norwood Venture Corp.
Mark R. Littell, President
145 West 45th Street, Suite 1211
New York, NY 10036
(212) 869-5075

Onondaga Venture Capital
Fund, Inc.
Irving W. Schwartz, Exec. V.P.
327 State Tower Building
Syracuse, NY 13202
(315) 478-0157

Preferential Capital Corporation
Bruce Bayroff, Secretary-
Treasurer
16 Court Street
Brooklyn, NY 11241
(718) 855-2728

Pyramid Ventures, Inc.
John Popovitch, Treasurer
280 Park Avenue, 10 West
New York, NY 10015
(212) 850-1934

Questech Capital Corp.
John E. Koonce, President
320 Park Avenue, 3rd Floor
New York, NY 10022
(212) 891-7500

R & R Financial Corp.
Imre Rosenthal, President
1451 Broadway
New York, NY 10036
(212) 790-1441

Rand SBIC, Inc.
Donald Ross, President
1300 Rand Building
Buffalo, NY 14203
(716) 853-0802

Realty Growth Capital
Corporation
Alan Leavit, President
271 Madison Avenue
New York, NY 10016
(212) 983-6880

Republic SBI Corporation
Robert V. Treanor, Senior V.P.
452 Fifth Avenue
New York, NY 10018
(212) 930-8639

SLK Capital Corp.
Edward A. Kerbs, President
115 Broadway, 20th Floor
New York, NY 10006
(212) 587-8800

Small Bus. Electronics
Investment Corp.
Stanley Meisels, President
1220 Peninsula Blvd.
Hewlett, NY 11557
(516) 374-0743

Southern Tier Capital
Corporation
Harold Gold, Secretary-
Treasurer
55 South Main Street
Liberty, NY 12754
(914) 292-3030

Sterling Commercial Capital, Inc.
Harvey L. Granat, President
175 Great Neck Road, Suite 404
Great Neck, NY 11021
(516) 482-7374

TLC Funding Corp.
Philip G. Kass, President
141 South Central Avenue
Hartsdale, NY 10530
(914) 683-1144

Tappan Zee Capital Corp.
(Main Office: Little Falls, NJ)
120 North Main Street
New City, NY 10956
(914) 634-8890

Telesciences Capital Corporation
Mike A. Petrozzo, Contact
26 Broadway, Suite 841
New York, NY 10004
(212) 425-0320

Vega Capital Corp.
Victor Harz, President
720 White Plains Road
Scarsdale, NY 10583
(914) 472-8550

Venture SBIC, Inc.
Arnold Feldman, President
249-12 Jericho Turnpike
Floral Park, NY 11001
(516) 352-0068

WFG-Harvest Partners, Ltd.
Harvey J. Wertheim, General
Partner
767 Third Avenue
New York, NY 10017
(212) 838-7776

Winfield Capital Corp.
Stanley M. Pechman, President
237 Mamaroneck Avenue
White Plains, NY 10605
(914) 949-2600

Wood River Capital Corporation
Thomas A. Barron, President
667 Madison Avenue
New York, NY 10022
(212) 750-9420

North Carolina

Delta Capital, Inc.
Alex B. Wilkins, Jr., President
227 N. Tryon Street, Suite 201
Charlotte, NC 28202
(704) 372-1410

Falcon Capital Corp.
P.S. Prasad, President
400 West Fifth Street
Greenville, NC 27834
(919) 752-5918

Heritage Capital Corp.
William R. Starnes, President
2095 Two First Union Center
Charlotte, NC 28282
(704) 334-2867

Kitty Hawk Capital, L.P.
Walter H. Wilkinson, President
Independence Center, Suite 1640
Charlotte, NC 28246
(704) 333-3777

NCNB SBIC Corporation
Troy S. McCrory, Jr., President
One NCNB Plaza—T05—2
Charlotte, NC 28255
(704) 374-5583

NCNB Venture Company, L.P.
S. Epes Robinson, General
Partner
One NCNB Plaza, T-39
Charlotte, NC 28255
(704) 374-5723

Ohio

A.T. Capital Corp.
Robert C. Salipante, President
900 Euclid Avenue, T-18
Mail: P.O. Box 5937
Cleveland, OH 44101
(216) 687-4970

Capital Funds Corp.
Carl G. Nelson, Chief Inv.
Officer
800 Superior Avenue
Cleveland, OH 44114
(216) 344-5774

Clarion Capital Corp.
Morton A. Cohen, President
35555 Curtis Blvd.
Eastlake, OH 44094
(216) 953-0555

First Ohio Capital Corporation
David J. McMacken, General
Manager
606 Madison Avenue
Mail: P.O. Box 2061; Toledo, OH
43603
Toledo, OH 43604
(419) 259-7146

Gries Investment Company
Robert D. Gries, President
1500 Statler Office Tower
Cleveland, OH 44115
(216) 861-1146

JRM Capital Corp.
H.F. Meyer, President
110 West Streetsboro Street
Hudson, OH 44236
(216) 656-4010

National City Capital
Corporation
Michael Sherwin, President
629 Euclid Avenue
Cleveland, OH 44114
(216) 575-2491

SeaGate Venture Management,
Inc.
Charles A. Brown, Vice-
President
245 Summit Street, Suite 1403
Toledo, OH 43603
(419) 259-8605

Tamco Investors (SBIC)
Incorporated
Nathan H. Monus, President
375 Victoria Road
Youngstown, OH 44515
(216) 792-3811

Oklahoma

Alliance Business Investment
Company
Barry Davis, President
17 East Second Street
One Williams Center, Suite 2000
Tulsa, OK 74172
(918) 584-3581

Western Venture Capital
Corporation
William B Baker, Chief
Operating Officer
4880 South Lewis
Tulsa, OK 74105
(918) 749-7981

Oregon

First Interstate Capital, Inc.
(Main Office: Newport Beach,
CA)
227 S.W. Pine Street, Suite 200
Portland, OR 97204
(503) 223-4334

Northern Pacific Capital Corp.
John J. Tennant, Jr., President
1201 S.W. 12th Avenue, Suite
608
Mail: P.O. Box 1658; Portland,
OR 97207
Portland, OR 97205
(503) 241-1255

Norwest Growth Fund, Inc.
(Main Office: Minneapolis, MN)
1300 S.W. 5th Street, Suite 3108
Portland, OR 97201
(503) 223-6622

U.S. Bancorp Capital Corp.
Stephen D. Fekety, President
111 S.W. 5th Avenue, Suite 1570
Portland, OR 97204
(503) 275-5860

Pennsylvania

Capital Corporation of America
Martin M. Newman, President
225 South 15th Street, Suite 920
Philadelphia, PA 19102
(215) 732-1666

Enterprise Venture Cap. Corp.
of Pennsylvania
Don Cowie, C.E.O.
227 Franklin Street, Suite 215
Johnstown, PA 15901
(814) 535-7597

Erie SBIC
George R. Heaton, President
32 West 8th Street, Suite 615
Erie, PA 16501
(814) 453-7964

Fidelcor Capital Corporation
Bruce H. Luehrs, President
123 S. Broad Street
Philadelphia, PA 19109
(215) 985-7287

First SBIC of California
(Main Office: Costa Mesa, CA)
Daniel A. Dye, Contact
P.O. Box 512
Washington, PA 15301
(412) 223-0707

First Valley Capital Corporation
Matthew W. Thomas, President
640 Hamilton Mall, 8th Floor
Allentown, PA 18101
(215) 776-6760

Meridian Capital Corp.
Joseph E. Laky, President
Suite 222, Blue Bell West
650 Skippack Pike
Blue Bell, PA 19422
(215) 278-8907

Meridian Venture Partners
Raymond R. Rafferty, General
Partner
The Fidelity Court Building
259 Radnor-Chester Road
Radnor, PA 19087
(215) 293-0210

PNC Capital Corp.
Gary J. Zentner, President
Pittsburgh National Building
Fifth Avenue and Wood Street
Pittsburgh, PA 15222
(412) 355-2245

Rhode Island

Domestic Capital Corp.
Nathaniel B. Baker, President
815 Reservoir Avenue
Cranston, RI 02910
(401) 946-3310

Fleet Venture Resources, Inc.
Robert M. Van Degna, President
111 Westminster Street
Providence, RI 02903
(401) 278-6770

Moneta Capital Corp.
Arnold Kilberg, President
285 Governor Street
Providence, RI 02906
(401) 861-4600

Old Stone Capital Corporation
Arthur C. Barton, President
One Old Stone Square, 11th
Floor
Providence, RI 02903
(401) 278-2559

Wallace Capital Corporation
Lloyd W. Granoff, President
170 Westminister Street
Suite 300
Providence, RI 02903
(401) 273-9191

South Carolina

Carolina Venture Cap. Corp.
Thomas H. Harvey III, President
14 Archer Road
Hilton Head Isl., SC 29928
(803) 842-3101

Charleston Capital Corporation
Henry Yaschik, President
111 Church Street
P.O. Box 328
Charleston, SC 29402
(803) 723-6464

Floco Investment Company, Inc.
(The)
William H. Johnson, Sr.,
President
Highway 52 North
Mail: P.O. Box 919; Lake City,
SC 29560
Scranton, SC 29561
(803) 389-2731

Lowcountry Investment
Corporation
Joseph T. Newton, Jr., President
4444 Daley Street
P.O. Box 10447
Charleston, SC 29411
(803) 554-9880

Reedy River Ventures
John M. Sterling, President
233 East Main Street, Suite 202
Mail: P.O. Box 17526
Greenville, SC 29606
(803) 232-6198

Tennessee

Financial Resources,
Incorporated
Milton Picard, Chairman of the
Board
2800 Sterick Building
Memphis, TN 38103
(901) 527-9411

Leader Capital Corp.
James E. Pruitt Jr., President
158 Madison Avenue
P.O. Box 708; Memphis, TN
38101-0708
Memphis, TN 38101
(901) 578-2405

Texas

Alliance Business Investment
Company
(Main Office: Tulsa, OK)
911 Louisiana
One Shell Plaza, Suite 3990
Houston, TX 77002
(713) 224-8224

Brittany Capital Company
Steve Peden, Partner
1525 Elm Street
2424 LTV Tower
Dallas, TX 75201
(214) 954-1515

Business Capital Corp.
James E. Sowell, Chairman of the
Board
4809 Cole Avenue, Suite 250
Dallas, TX 75205
(214) 522-3739

Capital Marketing Corporation
Ray Ballard, Manager
100 Nat Gibbs Drive
P.O. Box 1000
Keller, TX 76248
(817) 656-7309

Capital Southwest Venture Corp.
William R. Thomas, President
12900 Preston Road, Suite 700
Dallas, TX 75230
(214) 233-8242

Central Texas SBI Corporation
David G. Horner, President
P.O. Box 2600
Waco, TX 76702
(817) 753-6461

Charter Venture Group,
Incorporated
Winston C. Davis, President
2600 Citadel Plaza Drive, Suite
600
Houston, TX 77008
(713) 863-0704

Citicorp Venture Capital, Ltd.
(Main Office: New York, NY)
717 North Harwood
Suite 2920-LB87
Dallas, TX 75201
(214) 880-9670

Energy Assets, Inc.
Laurence E. Simmons, Exec. V.P.
4900 Republic Bank Center
700 Louisiana
Houston, TX 77002
(713) 236-9999

Enterprise Capital Corporation
Fred Zeidman, President
4543 Post Oak Place, #130
Houston, TX 77027
(713) 621-9444

FCA Investment Company
Robert S. Baker, Chairman
3000 Post Oak, Suite 1790
Houston, TX 77056
(713) 965-0061

First Interstate Cap. Corp. of
Texas
Richard S. Smith, President
1000 Louisiana, 7th Floor
Mail: P.O. Box 3326; Houston,
TX 77253
Houston, TX 77002
(713) 224-6611

Ford Capital, Ltd.
C. Jeff Pan, President
1525 Elm Street
Mail: P.O. Box 2140; Dallas, TX
75221
Dallas, TX 75201
(214) 954-0688

Houston Partners, SBIP
Harvard Hill, President, CGP
Capital Center Penthouse
401 Louisiana
Houston, TX 77002
(713) 222-8600

MCap Corp.
J. Wayne Gaylord, Manager
1717 Main Street, 6th Floor
Momentum Place
Dallas, TX 75201
(214) 939-3131

MVenture Corp
Wayne Gaylord, President
1717 Main Street, 6th Fl.
Momentum Place (Mail: P.O.
Box 662090; Dallas, TX 75266)
Dallas, TX 75201
(214) 939-3131

Mapleleaf Capital Ltd.
Edward Fink, President
55 Waugh, Suite 710
Houston, TX 77007
(713) 880-4494

Mid-State Capital Corporation
Smith E. Thomasson, President
510 North Valley Mills Drive
Waco, TX 76710
(817) 772-9220

Neptune Capital Corporation
Richard C. Strauss, President
5956 Sherry Lane, Suite 800
Dallas, TX 75225
(214) 739-1414

Omega Capital Corporation
Theodric E. Moor, Jr., President
755 South 11th Street, Suite 250
Mail: P.O. Box 2173
Beaumont, TX 77704
(409) 832-0221

Republic Venture Group,
Incorporated
Robert H. Wellborn, CEO
325 N. St Paul 2829 Tower II
Mail: P.O. Box 655961; Dallas,
TX 75265
Dallas, TX 75201
(214) 922-3500

Revelation Resources, Ltd.
Mr. Chris J. Mathews, Manager
2929 Allen Parkway, Suite 1705
Houston, TX 77019
(713) 526-5623

Rust Capital Limited
Jack A. Morgan, Partner
114 West 7th Street, Suite 500
Austin, TX 78701
(512) 482-0806

SBI Capital Corp.
William E. Wright, President
6305 Beverly Hill Lane
Mail: P.O. Box 570368; Houston,
TX 77257
Houston, TX 77057
(713) 975-1188

San Antonio Venture Group,
Inc.
Domingo Bueno, President
2300 West Commerce Street
San Antonio, TX 78207
(512) 223-3633

South Texas SBIC
Kenneth L. Vickers, President
120 South Main Street
P.O. Box 1698
Victoria, TX 77902
(512) 573-5151

SW'n Venture Cap. of Texas, Inc.
James A. Bettersworth, President
1336 East Court Street
P.O. Box 1719
Seguin, TX 78155
(512) 379-0380

SW'n Venture Cap. of Texas, Inc.
(Main Office: Seguin, TX)
1250 N.E. Loop 410, Suite 300
San Antonio, TX 78209
(512) 822-9949

Sunwestern Capital Corporation
Thomas W. Wright, President
3 Forest Plaza
12221 Merit Drive, Suite 1300
Dallas, TX 75251
(214) 239-5650

Sunwestern Ventures, Ltd.
Thomas W. Wright, President
3 Forest Plaza
12221 Merit Drive, Suite 1300
Dallas, TX 75251
(214) 239-5650

Texas Commerce Investment Co.
Fred Lummis, Vice President
Texas Commerce Bank Bldg.,
30th Floor
712 Main Street
Houston, TX 77002
(713) 236-4719

UNCO Ventures, Inc.
John Gatti, President
909 Fannin Street, 7th Floor
Houston, TX 77010
(713) 853-2422

Wesbanc Ventures, Ltd.
Stuart Schube, General Partner
520 Post Oak Blvd., Suite 130
Houston, TX 77027
(713) 622-9595

Western Financial Capital Corp.
Mrs. Marion Rosemore, Pres.
17772 Preston Road, Suite 101
Dallas, TX 75252
(214) 380-0044

Vermont

Queneska Capital Corporation
Albert W. Coffrin, III, President
123 Church Street
Burlington, VT 05401
(802) 865-1806

Virginia

Crestar Capital
A. Hugh Ewing, III Managing
G.P.
9 South 12th Street, Third Floor
Richmond, VA 23219
(804) 643-7358

James River Capital Associates
A. Hugh Ewing, Managing Part.
9 South 12th Street
Mail: P.O. Box 1776; Richmond,
VA 23219
Richmond, VA 23214
(804) 643-7323

Metropolitan Capital Corp.
John B. Toomey, President
2550 Huntington Avenue
Alexandria, VA 22303
(703) 960-4698

Sovran Funding Corp.
David A. King, Jr., President
Sovran Center, 6th Floor
One Commercial Plaza; Mail:
P.O. Box 600
Norfolk, VA 23510
(804) 441-4041

Tidewater SBI Corp.
Gregory H. Wingfield, President
1214 First Virginia Bank Tower
101 St. Paul's Blvd.
Norfolk, VA 23510
(804) 627-2315

Washington

Capital Resource Corporation
T. Evans Wyckoff, President
1001 Logan Building
Seattle, WA 98101
(206) 623-6550

Northwest Business Investment
Corp.
C. Paul Sandifur, President
929 West Sprague Avenue
Spokane, WA 99204
(509) 838-3111

Seafirst Capital Corporation
David R. West, Exec. Vice Pres.
Columbia Seafirst Center
701 Fifth Avenue, P.O. Box
34103
Seattle, WA 98124
(206) 358-7441

U.S. Bancorp Capital
Corporation
(Main Office: Portland, OR)
1415 Fifth Avenue
Seattle, WA 98171
(206) 344-8105

Washington Trust Equity Corp.
John M. Snead, President
Washington Trust Financial
Center
P.O. Box 2127
Spokane, WA 99210
(509) 455-3821

Wisconsin

Banc One Venture Corp.
H. Wayne Foreman, President
111 East Wisconsin Avenue
Milwaukee, WI 53202
(414) 765-2274

Bando-McGlocklin Capital Corp.
George Schonath, Investment
Advisor
13555 Bishops Court, Suite 225
Brookfield, WI 53005
(414) 784-9010

Capital Investments, Inc.
Robert L. Banner, Vice Pres.
Commerce Building, Suite 400
744 North Fourth Street
Milwaukee, WI 53203
(414) 273-6560

M & I Ventures Corp.
John T. Byrnes, President
770 North Water Street
Milwaukee, WI 53202
(414) 765-7910

MorAmerica Capital Corp.
(Main Office: Cedar Rapids,
Iowa)
600 East Mason Street
Milwaukee, WI 53202
(414) 276-3839

Super Market Investors, Inc.
David H. Maass, President
23000 Roundy Drive
Mail: P.O. Box 473; Milwaukee
53202
Pewaukee, WI 53072
(414) 547-7999

Wisconsin Community Capital,
Inc.
Paul J. Eble, President
1 S. Pinckney Street, Suite 500
Madison, WI 53703
(608) 256-3441

301 (D) SBICS SPECIALIZING IN MINORITY FUNDING

Alabama

Alabama Capital Corporation
David C. Delaney, President
16 Midtown Park East
Mobile, AL 36606
(205) 476-0700

Alabama Small Business
Investment Company
Harold Gilchrist, Manager
206 North 24th Street
Birmingham, AL 35203
(205) 324-5234

Tuskegee Capital Corporation
A. G. Bartholomew, President
4453 Richardson Road
Hampton Hall Building
Montgomery, AL 36108
(205) 281-8059

Alaska

Calista Business Investment
Corp.
Alex Raider, President
503 East Sixth Avenue
Anchorage, AK 99501
(907) 277-0425

Arkansas

Capital Management Services,
Inc.
David L. Hale, President
1910 N. Grant Street, Suite 200
Little Rock, AR 72207
(501) 664-8613

Power Ventures, Inc.
Dorsey D. Glover, President
829 Highway 270 North
Malvern, AR 72104
(501) 332-3695

California

ABC Capital Corp.
Anne B. Cheng, President
610 East Live Oak Avenue
Arcadia, CA 91006
(818) 570-0653

Allied Business Investors, Inc.
Jack Hong, President
428 S. Atlantic Blvd., Suite 201
Monterey Park, CA 91754
(818) 289-0186

Ally Finance Corp.
Percy P. Lin, President
9100 Wilshire Blvd., Suite 408
Beverly Hills, CA 90212
(213) 550-8100

Asian American Capital Corp.
David Der, President
1251 West Tennyson Road
Suite #4
Hayward, CA 94544
(415) 887-6888

Astar Capital Corp.
George Hsu, President
7282 Orangethorpe Ave., Suite 8
Buena Park, CA 90621
(714) 739-2218

Bentley Capital
John Hung, President
592 Vallejo Street, Suite #2
San Francisco, CA 94133
(415) 362-2868

Best Finance Corporation
Vincent Lee, General Manager
1814 W. Washington Blvd.
Los Angeles, CA 90007
(213) 731-2268

Business Equity and
Development Corp.
Leon M.N. Garcia,
President/CEO
767 North Hill Street, Suite 401
Los Angeles, CA 90012
(213) 613-0916

Calsafe Capital Corp.
Bob T.C. Chang, Chairman of the
Board
240 South Atlantic Blvd.
Alhambra, CA 91801
(818) 289-4080

Charterway Investment
Corporation
Harold H. M. Chuang, President
222 South Hill Street, Suite 800
Los Angeles, CA 90012
(213) 687-8539

Continental Investors, Inc.
Lac Thantrong, President
8781 Seaspray Drive
Huntington Beach, CA 92646
(714) 964-5207

Equitable Capital Corporation
John C. Lee, President
855 Sansome Street
San Francisco, CA 94111
(415) 434-4114

First American Capital Funding,
Inc.
Luu TranKiem, Chairman
38 Corporate Park, Suite B
Irvine, CA 92714
(714) 660-9288

Helio Capital, Inc.
Chester Koo, President
5900 South Eastern Avenue
Suite 136
Commerce, CA 90040
(213) 721-8053

LaiLai Capital Corp.
Hsing-Jong Duan, Pres. &
General Manager
223 E. Garvey Avenue, Suite 228
Monterey, CA 91754
(818) 288-0704

Magna Pacific Investments
David Wong, President
700 N. Central Avenue, Suite 245
Glendale, CA 91203
(818) 547-0809

Myriad Capital, Inc.
Felix Chen, President
328 S. Atlantic Blvd., Suite 200 A
Monterey Park, CA 91754
(818) 570-4548

New Kukje Investment Company
George Su Chey, President
3670 Wilshire Blvd., Suite 418
Los Angeles, CA 90010
(213) 389-8679

Opportunity Capital Corporation
J. Peter Thompson, President
One Fremont Place
39650 Liberty Street, Suite 425
Fremont, CA 94538
(415) 651-4412

Positive Enterprises, Inc.
Kwok Szeto, President
399 Arguello Street
San Francisco, CA 94118
(415) 386-6606

RSC Financial Corp.
Frederick K. Bae, President
323 E. Matilija Road, #208
Ojai, CA 93023
(805) 646-2925

San Joaquin Business Investment
Group, Inc.
Joe Williams, President
2310 Tulare Street, Suite 140
Fresno, CA 93721
(209) 233-3580

Colorado

Colorado Invesco, Inc.
1999 Broadway, Suite 2100
Denver, CO 80202
(303) 293-2431

District of Columbia

Allied Financial Corporation
David J. Gladstone, President
1666 K Street, N.W., Suite 901
Washington, DC 20006
(202) 331-1112

Broadcast Capital, Inc.
John E. Oxendine, President
1771 N Street, N.W., Suite 421
Washington, DC 20036
(202) 429-5393

Consumers United Capital
Corporation
Ester M. Carr-Davis, President
2100 M Street, N.W.
Washington, DC 20037
(202) 872-5274

Fulcrum Venture Capital
Corporation
C. Robert Kemp, Chairman
1030 15th Street, N.W., Suite 203
Washington, DC 20005
(202) 785-4253

Minority Broadcast Investment
Corp.
Walter L. Threadgill, President
1200 18th Street, N.W.
Suite 705
Washington, DC 20036
(202) 293-1166

Syncom Capital Corp.
Herbert P. Wilkins, President
1030—15th Street, N.W., Suite
203
Washington, DC 20005
(202) 293-9428

Florida

Allied Financial Corporation
(Main Office: Washington, D.C.)
Executive Office Center, Suite
305
2770 N. Indian River Blvd.
Vero Beach, FL 32960
(407) 778-5556

First American Lending
Corporation (The)
Roy W. Talmo, Chairman
1926 10th Avenue North
Mail: P.O. Box 24660; W. Palm
Beach 33416
Lake Worth, FL 33461
(305) 533-1511

Ideal Financial Corporation
Ectore T. Reynaldo, General
Manager
780 N.W. 42nd Avenue, Suite
303
Miami, FL 33126
(305) 442-4665

Pro-Med Investment
Corporation
(Main Office: Dallas, TX)
1380 N.E. Miami Gardens Drive
Suite 225
N. Miami Beach, FL 33179
(305) 949-5900

Venture Group, Inc.
Ellis W. Hitzing, President
5433 Buffalo Avenue
Jacksonville, FL 32208
(904) 353-7313

Georgia

Renaissance Capital Corporation
Samuel B. Florence, President
161 Spring Street, NW
Suite 610
Atlanta, GA 30303
(404) 658-9061

Hawaii

Pacific Venture Capital, Ltd.
Dexter J. Taniguchi, President
222 South Vineyard Street
PH.l
Honolulu, HI 96813
(808) 521-6502

Illinois

Amoco Venture Capital Co.
Gordon E. Stone, President
200 E. Randolph Drive
Chicago, IL 60601
(312) 856-6523

Chicago Community Ventures,
Inc.
Phyllis George, President
104 South Michigan Avenue
Suite 215-218
Chicago, IL 60603
(312) 726-6084

Combined Fund, Inc. (The)
E. Patric Jones, President
1525 East 53rd Street
Chicago, IL 60615
(312) 753-9650

Neighborhood Fund, Inc. (The)
James Fletcher, President
1950 East 71st Street
Chicago, IL 60649
(312) 684-8074

Peterson Finance and Investment
Company
James S. Rhee, President
3300 West Peterson Avenue,
Suite A
Chicago, IL 60659
(312) 583-6300

Tower Ventures, Inc.
Robert T. Smith, President
Sears Tower, BSC 43-50
Chicago, IL 60684
(312) 875-0571

Kentucky

Equal Opportunity Finance, Inc.
Franklin Justice, Jr., V.P. &
Manager
420 Hurstbourne Lane, Suite 201
Louisville, KY 40222
(502) 423-1943

Louisiana

SCDF Investment Corp.
Martial Mirabeau, Manager
1006 Surrey Street
P.O. Box 3885
Lafayette, LA 70502
(318) 232-3769

Maryland

Albright Venture Capital, Inc.
William A. Albright, President
1355 Piccard Drive, Suite 380
Rockville, MD 20850
(301) 921-9090

Security Financial and
Investment Corp.
Han Y. Cho, President
7720 Wisconsin Avenue, Suite
207
Bethesda, MD 20814
(301) 951-4288

Massachusetts

Argonauts MESBIC Corporation
(The)
Mr. Chi Fu Yeh, President
2 Vernon Street
P.O. Box 2411
Framingham, MA 01701
(508) 820-3430

New England MESBIC, Inc.
Etang Chen, President
530 Turnpike Street
North Andover, MA 01845
(617) 688-4326

Transportation Capital Corp.
(Main Office: New York, NY)
45 Newbury Street, Suite 207
Boston, MA 02116
(617) 536-0344

Michigan

Dearborn Capital Corp.
Michael LaManes, President
P.O. Box 1729
Dearborn, MI 48121
(313) 337-8577

Metro-Detroit Investment Co.
William J. Fowler, President
30777 Northwestern Highway,
Suite 300
Farmington Hills, MI 48018
(313) 851-6300

Motor Enterprises, Inc.
James Kobus, Manager
3044 West Grand Blvd.
Detroit, MI 48202
(313) 556-4273

Mutual Investment Co., Inc.
Jack Najor, President
21415 Civic Center Drive
Mark Plaza Building, Suite 217
Southfield, MI 48076
(313) 557-2020

Minnesota

Capital Dimensions Ventures
Fund, Inc.
Dean R. Pickerell, President
Two Appletree Square, Suite 244
Minneapolis, MN 55425
(612) 854-3007

Mississippi

Sun-Delta Capital Access Center,
Inc.
Howard Boutte, Jr., Vice Pres.
819 Main Street
Greenville, MS 38701
(601) 335-5291

New Jersey

Capital Circulation Corporation
Judy Kao, Manager
208 Main Street
Fort Lee, NJ 07024
(201) 947-8637

Formosa Capital Corp.
Philp Chen, President
1037 Route 46 East, Unit C-208
Clifton, NJ 07013
(201) 916-0016

Rutgers Minority Investment
Company
Oscar Figueroa, President
92 New Street
Newark, NJ 07102
(201) 648-5287

Transpac Capital Corporation
Tsuey Tang Wang, President
1037 Route 46 East
Clifton, NJ 07013
(201) 470-0706

Zaitech Capital Corporation
Mr. Fu-Tong Hsu, President
1037 Route 46 East, Unit C-201
Clifton, NJ 07013
(201) 365-0047

New Mexico

Associated Southwest Investors,
Inc.
John R. Rice, General Manager
2400 Louisiana N.E.
Bldg. #4, Suite 225
Albuquerque, NM 87110
(505) 881-0066

New York

American Asian Capital Corp.
Howard H. Lin, President
130 Water Street, Suite 6-L
New York, NY 10005
(212) 422-6880

Avdon Capital Corp.
A. M. Donner, President
1413 Avenue J
Brooklyn, NY 11230
(718) 692-0950

CVC Capital Corp.
Jeorg G. Klebe, President
131 East 62th Street
New York, NY 10021
(212) 319-7210

Capital Investors & Management
Corp.
Rose Chao, Manager
210 Canal Street, Suite 607
New York, NY 10013
(212) 964-2480

Cohen Capital Corp.
Edward H. Cohen, President
8 Freer Street, Suite 185
Lynbrook, NY 11563
(516) 887-3434

Columbia Capital Corporation
Mark Scharfman, President
79 Madison Avenue, Suite 800
New York, NY 10016
(212) 696-4334

East Coast Venture Capital, Inc.
Zindel Zelmanovitch, President
313 West 53rd Street, Third
Floor
New York, NY 10019
(212) 245-6460

Elk Associates Funding
Corporation
Gary C. Granoff, President
600 Third Avenue, 38th Floor
New York, NY 10016
(212) 972-8550

Equico Capital Corp.
Duane Hill, President
135 West 50th Street, 11th Floor
New York, NY 10020
(212) 641-7650

Everlast Capital Corporation
Frank J. Segreto, Gen. Mgr. &
V.P.
350 Fifth Avenue, Suite 2805
New York, NY 10118
(212) 695-3910

Exim Capital Corp.
Victor K. Chun, President
290 Madison Avenue
New York, NY 10017
(212) 683-3375

Fair Capital Corp.
Robert Yet Sen Chen, President
c/o Summit Associates
3 Pell Street, 2nd Floor
New York, NY 10013
(212) 608-5866

Freshstart Venture Capital Corp.
Zindel Zelmanovich, President
313 West 53rd Street, 3rd Floor
New York, NY 10019
(212) 265-2249

Hanam Capital Corp.
Dr. Yul Chang, President
One Penn Plaza, Ground Floor
New York, NY 10119
(212) 714-9830

Hop Chung Capital Investors,
Inc.
Yon Hon Lee, President
185 Canal Street, Room 303
New York, NY 10013
(212) 219-1777

Horn & Hardart Capital Corp.
Gerald Zarin, Vice President
730 Fifth Avenue
New York, NY 10019
(212) 484-9600

Ibero American Investors Corp.
Emilio Serrano, President
38 Scio Street
Rochester, NY 14604
(716) 262-3440

Intercontinental Capital Funding Corp.
James S. Yu, President
60 East 42nd Street, Suite 740
New York, NY 10165
(212) 286-9642

International Paper Cap.
Formation, Inc.
(Main Office: Memphis, TN)
Frank Polney, Manager
Two Manhattanville Road
Purchase, NY 10577
(914) 397-1578

Japanese American Capital Corp.
Stephen C. Huang, President
19 Rector Street
New York, NY 10006
(212) 964-4077

Jardine Capital Corp.
Evelyn Sy Dy, President
109 Lafayette Street, Unit 204
New York, NY 10038
(212) 941-0966

Manhattan Central Capital Corp.
David Choi, President
1255 Broadway, Room 405
New York, NY 10001
(212) 684-6411

Medallion Funding Corporation
Alvin Murstein, President
205 E. 42nd Street, Suite 2020
New York, NY 10017
(212) 682-3300

Minority Equity Cap. Co., Inc.
Donald F. Greene, President
275 Madison Avenue
New York, NY 10016
(212) 686-9710

Monsey Capital Corp.
Shamuel Myski, President
125 Route 59
Monsey, NY 10952
(914) 425-2229

New Oasis Capital Corporation
James Huang, President
114 Liberty Street, Suite 304
New York, NY 10006
(212) 349-2804

North American Funding Corporation
Franklin F. Y. Wong, V.P. & Gen. Mgr.
177 Canal Street
New York, NY 10013
(212) 226-0080

Pan Pac Capital Corp.
Dr. In Ping Jack Lee, President
121 East Industry Court
Deer Park, NY 11729
(516) 586-7653

Pierre Funding Corp.
605 Third Avenue
New York, NY 10016
(212) 490-9540

Situation Venture Corporation
Sam Hollander, President
502 Flushing Avenue
Brooklyn, NY 11205
(718) 855-1835

Square Deal Venture Capital Corp.
Mordechai Z. Feldman, President
805 Avenue L
Brooklyn, NY 11230
(718) 692-2924

Taroco Capital Corp.
David R. C. Chang, President
19 Rector Street, 35th Floor
New York, NY 10006
(212) 344-6690

Transportation Capital Corp.
Melvin L. Hirsch, President
60 East 42nd Street, Suite 3115
New York, NY 10165
(212) 697-4885

Triad Capital Corp. of New York
Lorenzo J. Barrera, President
960 Southern Blvd.
Bronx, NY 10459
(212) 589-6541

Trico Venture, Inc.
Avruhum Donner, President
1413 Avenue J
Brooklyn, NY 11230
(718) 692-0950

United Capital Investment Corp.
Paul Lee, President
60 East 42nd Street, Suite 1515
New York, NY 10165
(212) 682-7210

Venture Opportunities Corp.
A. Fred March, President
110 East 59th Street, 29th Floor
New York, NY 10022
(212) 832-3737

Watchung Capital Corp.
S. T. Jeng, President
431 Fifth Avenue, Fifth Floor
New York, NY 10016
(212) 889-3466

Yang Capital Corp.
Maysing Yang, President
41-40 Kissena Blvd.
Flushing, NY 11355
(516) 482-1578

Yusa Capital Corp.
Christopher Yeung, Chairman of the Board
622 Broadway
New York, NY 10012
(212) 420-1350

North Carolina

Business Capital Inv. Co., Inc.
Christopher S. Liu, Manager
327 South Road
High Point, NC 27260
(919) 889-8334

Ohio

Center City MESBIC, Inc.
Michael A. Robinson, President
Centre City Office Building,
Suite 762
40 South Main Street
Dayton, OH 45402
(513) 461-6164

Rubber City Capital Corporation
Jesse T. Williams, President
1144 East Market Street
Akron, OH 44316
(216) 796-9167

Pennsylvania

Alliance Enterprise Corporation
W. B. Priestley, President
1801 Market Street, 3rd Floor
Philadelphia, PA 19103
(215) 977-3925

Greater Phila. Venture Capital
Corp., Inc.
Martin Newman, Manager
920 Lewis Tower Bldg.
225 South Fifteenth Street
Philadelphia, PA 19102
(215) 732-3415

Salween Financial Services, Inc.
Dr. Ramarao Naidu, President
228 North Pottstown Pike
Exton, PA 19341
(215) 524-1880

Puerto Rico

North America Inv. Corporation
Santigo Ruz Betacourt, President
Banco CTR #1710, M Rivera Av
Stop 34
Mail: PO BX 1831 Hato Rey
Sta., PR 00919
Hato Rey, PR 00936
(809) 751-6178

Tennessee

Chickasaw Capital Corporation
Tom Moore, President
67 Madison Avenue
Memphis, TN 38147
(901) 523-6404

International Paper Cap.
Formation, Inc.
John G. Herman, V.P. and
Controller
International Place I
6400 Poplar Avenue, 10-74
Memphis, TN 38197
(901) 763-6282

Tennessee Equity Capital
Corporation
Walter S. Cohen, President
1102 Stonewall Jackson Court
Nashville, TN 37220
(615) 373-4502

Tennessee Venture Capital
Corporation
Wendell P. Knox, President
162 Fourth Avenue North, Suite
125
Mail: P.O. Box 2567
Nashville, TN 37219
(615) 244-6935

Valley Capital Corp.
Lamar J. Partridge, President
8th Floor Krystal Building
100 W. Martin Luther King Blvd.
Chattanooga, TN 37402
(615) 265-1557

West Tennessee Venture Capital
Corporation
Osbie L. Howard, President
152 Beale Street, Suite 401
Mail: P.O. Box 300; Memphis,
TN 38101
Memphis, TN 38101
(901) 527-6091

Texas

Chen's Financial Group, Inc.
Samuel S. C. Chen, President
1616 West Loop South, Suite 200
Houston, TX 77027
(713) 850-0879

Evergreen Capital Company, Inc.
Shen-Lim Lin, Chairman &
President
8502 Tybor Drive, Suite 201
Houston, TX 77074
(713) 778-9770

MESBIC Financial Corp. of
Dallas
Donald R. Lawhorne, President
12655 N. Central Expressway
Suite 814
Dallas, TX 75243
(214) 991-1597

MESBIC Financial Corp. of
Houston
Lynn H. Miller, President
811 Rusk, Suite 201
Houston, TX 77002
(713) 228-8321

Minority Enterprise Funding,
Inc.
Frederick C. Chang, President
17300 El Camino Real, Suite
107-B
Houston, TX 77058
(713) 488-4919

Pro-Med Investment Corp.
Mrs. Marion Rosemore, Pres.
17772 Preston Road, Suite 101
Dallas, TX 75252
(214) 380-0044

Southern Orient Capital
Corporation
Min H. Liang, President
2419 Fannin, Suite 200
Houston, TX 77002
(713) 225-3369

United Oriental Capital Corp.
Don J. Wang, President
908 Town & Country Blvd.
Suite 310
Houston, TX 77024
(713) 461-3909

Virginia

East West United Investment
Company
Bui Dung, President
815 West Broad Street
Falls Church, VA 22046
(703) 237-7200

Washington Finance and
Investment Corp.
Chang H. Lie, President
100 E. Broad Street
Falls Church, VA 22046
(703) 534-7200

Wisconsin

Future Value Ventures, Inc.
William P. Beckett, President
622 N. Water Street, Suite 500
Milwaukee, WI 53202
(414) 278-0377

SEED CAPITAL NETWORKS

The Computerized Ontario
Investment Network (COIN)
Ontario Chamber of Commerce
2323 Yonge Street
Toronto, Ontario, Canada M4P
2C9
(416) 482-5222
Region served: Ontario

Venture Capital Network of
Atlanta, Inc.
230 Peachtree Street N.E.
Suite 1810
Atlanta, GA 30303
(404) 658-7000
Region served: Georgia

Heartland Venture Capital
Network
Evanston Business Investment
Corp.
1710 Orrington Avenue
Evanston, IL 60201
(312) 864-7970
Region served: Illinois

Indiana Seed Capital Network
Institute of New Business
Ventures, Inc.
One North Capital, Suite 420
Indianapolis, IN 46204
(317) 634-8418
Region served: Indiana

Upper Peninsula Venture
Capital Network, Inc.
206 Cohodas Administration
Center
Northern Michigan University
Marquette, MI 49855
(906) 227-2406
Region served: Michigan

Venture Capital Network of
Minnesota
23 Empire Drive
St. Paul, MN 55103
(612) 223-8663
Region served: Minnesota

Midwest Venture Capital
Network
P.O. Box 4659
St. Louis, MO 63108
(314) 534-7204
Region served: Missouri

Mississippi Venture Capital
Clearinghouse
Mississippi Research and
Development Center
3825 Ridgewood Road
Jackson, MS 39211
(601) 982-6425
Region served: Mississippi

Investment Contact Network
Institute for the Study of Private
Enterprise
University of North Carolina
The Kenan Center 498A
Chapel Hill, NC 27514
(919) 962-8201
Region served: North Carolina

Venture Capital Network, Inc.
P.O. Box 882
Durham, NH 03824
(603) 862-3556
No geographic limit

Venture Capital Network of New
York, Inc.
TAC—State University
College of Arts and Science
Plattsburgh, NY 12901
(518) 564-2214
Region served: New York

Univ. of South Carolina at Aiken
171 University Parkway
Aiken, SC 29801
(803) 648-6851
Region served: South Carolina

Seed Capital Network, Inc.
Operations Center
8905 Kingston Pike, Suite 12493
Knoxville, TN 37923
(615) 693-2091
No geographic limit

Venture Capital Network of
Texas
P.O. Box 690870
San Antonio, TX 78269-0870
(512) 691-4318
Region served: Texas

Casper College
Small Business Development
Center
125 College Drive
Casper, WY 82601
(307) 235-4825
Region served: Wyoming

XI. SBA REGIONAL AND DISTRICT OFFICES

Address inquiries to "SBA" and the address given.

2121 8th Ave. N., Suite 200, Birmingham, AL, 35203-2398, (205) 731-1344

330 Main Street, 2nd Floor, Hartford, CT, 06106. (203) 240-4700

1111 18th Street NW, 6th Floor, Washington, DC, 20036. (202) 634-4950

844 King Street, Room 5207, Wilmington, DE, 19801. (302) 573-6294

1320 S. Dixie Highway, Suite 501, Coral Gables, FL, 33146. (305) 536-5521

400 W. Bay Street, Room 261, Jacksonville, FL, 32202. (904) 791-3782

700 Twiggs Street, Room 607, Tampa, FL, 33602 (813) 228-2594

5601 Corporate Way S., Suite 402, W. Palm Beach, FL, 33407. (407) 689-3922

1375 Peachtree St., NE, 5th Floor, Atlanta, GA, 30367-8102. (404) 347-2797

1720 Peachtree Rd., NW, 6th Floor, Atlanta, GA, 30309. (404) 347-2441

52 N. Main Street, Room 225, Statesboro, GA, 30458. (912) 489-8719

230 S. Dearborn Street, Room 510, Chicago, IL, 60604-1593. (312) 353-0359

219 S. Dearborn Street, Room 437, Chicago, IL, 60604-1779. (312) 353-4528

511 W. Capitol Street, Suite 302, Springfield, IL, 62704. (217) 492-4416

575 N. Pennsylvania St., Room 578, Indianapolis, IN, 46204-1584. (317) 269-7272

600 Federal Place, Room 188, Louisville, KY, 40202. (502) 582-5976

60 Batterymarch Street, 10th Floor, Boston, MA, 02110. (617) 451-2030

10 Causeway Street, Room 265, Boston, MA, 02114. (617) 565-5590

1550 Main Street, Room 212, Springfield, MA, 01103. (413) 785-0268

10 N. Calvert Street, 3rd Floor, Baltimore, MD, 21202. (301) 962-4392

40 Western Avenue, Room 512, Augusta, ME, 04330. (207) 622-8378

477 Michigan Ave., Room 515, Detroit, MI, 48226. (313) 226-6075

300 S. Front St., Marquette, MI, 49885. (906) 225-1108

100 N. 6th Street, Suite 610, Minneapolis, MN, 55403-1563. (612) 370-2324

One Hancock Plaza, Suite 1001, Gulfport, MS, 39501-7758. (601) 863-4449

100 W. Capitol Street, Suite 322, Jackson, MS, 39269-0396. (601) 965-4378

222 S. Church Street, Room 300, Charlotte, NC, 28202. (704) 371-6563

55 Pleasant Street, Room 210, Concord, NH, 03301-1257. (603) 225-1400

2600 Mt. Ephrain Ave., Camden, NJ, 08104. (609) 757-5183

60 Park Place, 4th Floor, Newark, NJ, 07102. (201) 645-2434

445 Broadway, Room 261, Albany, NY, 12207. (518) 472-6300

111 W. Huron Street, Room 1311, Buffalo, NY, 14202. (716) 846-4301

333 E. Water Street, 4th Floor, Elmira, NY, 14901. (607) 734-8130

35 Pinelawn Road, Room 102E, Melville, NY, 11747. (516) 454-0750

26 Federal Plaza, Room 3100, New York, NY, 10278. (212) 264-4355

26 Federal Plaza, Room 31-08, New York, NY, 10278. (212) 264-7772

100 State Street, Room 601, Rochester, NY, 14614. (716) 263-6700

100 S. Clinton Street, Room 1071, Syracuse, NY, 13260.
(315) 423-5383

550 Main Street, Room 5028, Cincinnati, OH, 45202.
(513) 684-2814

1240 E. 9th Street, Room 317, Cleveland, OH, 44199.
(216) 522-4180

85 Marconi Blvd., Room 512, Columbus, OH, 43215.
(614) 469-6860

100 Chestnut Street, Suite 309, Harrisburg, PA, 17101.
(717) 782-3840

475 Allendale Road, Suite 201, King of Prussia, PA, 19406.
(215) 962-3700

960 Penn Avenue, 5th Floor, Pittsburgh, PA, 15222.
(412) 644-2780

20 N. Pennsylvania Ave., Room 2327, Wilkes-Barre, PA, 18701.
(717) 826-6497

Carlos Chardon Ave., Room 691, Hato Rey, PR, 00918.
(809) 753-4002

380 Westminister Mall, 5th Floor, Providence, RI, 02903.
(401) 528-4586

1835 Assembly Street, Room 358, Columbia, SC, 29202.
(803) 765-5376

404 James Robertson Pkwy., Suite 1012, Nashville, TN, 37219.
(615) 736-5881

400 N. 8th Street, Room 3015, Richmond, VA, 23240.
(804) 771-2617

4C & 4D Este Sion Frm, Room 7, St. Croix, VI, 00820.
(809) 778-5380

Veterans Drive, Room 283, St. Thomas, VI, 00801.
(809) 774-8530

87 State Street, Room 205, Montpelier, VT, 05602.
(802) 828-4474

500 S. Barstow Commons, Room 37, Eau Claire, WI, 54701.
(715) 834-9012

212 E. Washington Ave., Room 213, Madison, WI, 53703.
(608) 264-5261

310 W. Wisconsin Ave., Suite 400, Milwaukee, WI, 53203.
(414) 291-3941

550 Eagan Street, Suite 309, Charleston, WV, 25301.
(304) 347-5220

168 W. Main Street, 5th Floor, Clarksburg, WV, 26301.
(304) 623-5631

XII. SBA LITERATURE FOR THE INVENTOR AND ENTREPRENEUR

FINANCIAL MANAGEMENT AND ANALYSIS

FM 1 ABC's OF BORROWING
Some small business people cannot understand why a lending institution refused to lend them money. Others have no trouble getting funds but are surprised to find strings attached to their loans. Learn the fundamentals of borrowing. . .$1.00.

FM 2 PROFIT COSTING AND PRICING FOR MANUFACTURERS
Uncover the latest techniques for pricing your product's profitably . . . $1.00.

FM 3 BASIC BUDGETS FOR PROFIT PLANNING
This publication takes the worry out of putting together a comprehensive budgeting system to monitor your profits and assess your financial operations . . . 50 cents.

FM 4 UNDERSTANDING CASH FLOW
In order to survive, a business must have enough cash to meet its obligations. This aid shows the owner/manager how to plan for the movement of cash through the business and thus plan for future requirements . . . $1.00.

FM 5 A VENTURE CAPITAL PRIMER FOR SMALL BUSINESS
This best-seller highlights the venture capital resources available and how to develop a pro

posal for obtaining these funds . . . 50 cents.

FM 6 ACCOUNTING SERVICES FOR SMALL SERVICE FIRMS
Sample profit/loss statements are used to illustrate how accounting services can help expose and correct trouble spots in a business' financial records . . . 50 cents.

FM 7 ANALYZE YOUR RECORDS TO REDUCE COSTS
Cost reduction IS NOT simply slashing any and all expenses. Understand the nature of expenses and how they inter-relate with sales, inventories and profits. Achieve greater profits through more efficient use of the dollar . . . 50 cents.

FM 8 BUDGETING IN A SMALL BUSINESS FIRM
Learn how to set up and keep sound financial records. Study how to effectively use journals, ledgers and charts to increase profits . . . 50 cents.

FM 9 SOUND CASH MANAGEMENT AND BORROWING
Avoid a "cash crisis" through proper use of cash budgets, cash flow projections and planned borrowing concepts . . . 50 cents.

FM 10 RECORD KEEPING IN A SMALL BUSINESS
Need some basic advice on setting up a useful record keeping system? This publication describes how . . . $1.00.

FM 11 BREAKEVEN ANALYSIS: A DECISION MAKING TOOL
Learn how "breakeven analysis" enables the manager/owner to make better decisions concerning sales, profits and costs . . . $1.00.

FM 13 PRICING YOUR PRODUCTS AND SERVICES PROFITABLY
Discusses how to price your products profitably, how to use the various techniques of pricing and when to use these techniques to your advantage . . . $1.00.

GENERAL MANAGEMENT AND PLANNING

MP 1 EFFECTIVE BUSINESS COMMUNICATIONS
Explains the importance of business communications and how they play a valuable role in business success . . . 50 cents.

MP 2 LOCATING OR RELOCATING YOUR BUSINESS
Learn how a company's market, available labor force, transportation and raw materials are affected when selecting a business location . . . $1.00.

MP 3 PROBLEMS IN MANAGING A FAMILY-OWNED BUSINESS
Specific problems exist when attempting to make family-owned businesses successful. This publication offers suggestions on how to overcome these difficulties . . . 50 cents.

MP 4 BUSINESS PLAN FOR SMALL MANUFACTURERS
Designed to help an owner/manager of a small manufacturing firm. This publication covers all the basic information necessary to develop an effective business plan . . . $1.00.

MP 6 PLANNING AND GOAL SETTING FOR SMALL BUSINESS
Learn how to plan for success . . . 50 cents.

MP 7 FIXING PRODUCTION MISTAKES
Structured as a checklist, this publication emphasizes the steps that should be taken by a manufacturer when a production mistake has been found . . . 50 cents.

MP 8 SHOULD YOU LEASE OR BUY EQUIPMENT?
Describes various aspects of the lease/buy decision. It lists advantages and disadvantages of leasing and provides a format for comparing the costs of the two . . . 50 cents.

MP 12 GOING INTO BUSINESS
This best-seller highlights important considerations you should know in reaching a decision to start your own business. It also includes a checklist for going into business . . . 50 cents.

MP 13 FEASIBILITY CHECKLIST FOR STARTING YOUR OWN BUSINESS
Helps you determine if your idea represents a real business opportunity. Assists in screening out ideas that are likely to fail, before you invest extensive time, money and effort in them . . . $1.00.

MP 14 HOW TO GET STARTED WITH A SMALL BUSINESS COMPUTER
Helps you forecast your computer needs, evaluate the alternative choices and select the right computer system for your business . . . $1.00.

MP 15 THE BUSINESS PLAN FOR HOMEBASED BUSINESS
Provides a comprehensive approach to developing a business plan for a homebased business. If you are convinced that a profitable home business is attainable, this publication will provide a step-by-step guide to develop a plan for your business . . . $1.00.

MP 16 HOW TO BUY OR SELL A BUSINESS
Learn several techniques used in determining the best price to buy or sell a small business . . . $1.00.

MP 17 PURCHASING FOR OWNERS OF SMALL PLANTS
Presents an outline of an effective purchasing program. Also includes a bibliography for further research into industrial purchasing . . . 50 cents.

MP 18 BUYING FOR RETAIL STORES
Discusses the latest trends in retail buying. Includes a bibliography that references a wide variety of private and public sources of information on most aspects of retail buying . . . $1.00.

MP 19 SMALL BUSINESS DECISION MAKING
Acquaint yourself with the wealth of information available on management approaches and tech

niques to identify, analyze and solve business problems . . . $1.00.

MP 20 BUSINESS CONTINUATION PLANNING
Provides an overview of business owners' life insurance needs that are not typically considered until after the death of one of the business' principal owners . . . $1.00.

MP 21 DEVELOPING A STRATEGIC BUSINESS PLAN
Helps you develop a formal strategic plan of action for your small business . . . $1.00.

MP 22 INVENTORY MANAGEMENT
Discusses the purpose of inventory management, types of inventories, record keeping and forecasting inventory levels . . . 50 cents.

MP 23 TECHNIQUES FOR PROBLEM SOLVING
Instructs the small business person on the key techniques of problem solving and problem identification, as well as designing and implementing a plan to correct these problems . . . $1.00.

MP 24 TECHNIQUES FOR PRODUCTIVITY IMPROVEMENT
Learn how to increase worker output through motivating "quality of work life" concepts and tailoring benefits to meet the needs of the employees . . . $1.00.

MP 25 SELECTING THE LEGAL STRUCTURE FOR YOUR BUSINESS
Discusses the various legal structures that a small business can use in setting up its operations. It briefly identifies the types of legal

structures and lists the advantages and disadvantages of each . . . 50 cents.

MP 26 EVALUATING FRANCHISE OPPORTUNITIES
Although the success rate for franchise-owned businesses is significantly better than start-up businesses, success is not guaranteed. Learn how to evaluate franchise opportunities and select the business that's right for you . . . 50 cents.

CRIME PREVENTION

CP 1 REDUCING SHOPLIFTING LOSSES
Learn the latest techniques on how to spot, deter, apprehend and prosecute shoplifters . . . 50 cents.

CP 2 CURTAILING CRIME – INSIDE AND OUT
Positive steps can be taken to curb crime. They include safeguards against employee dishonesty and ways to control shoplifting. In addition, this publication includes measures to outwit bad-check passing and ways to prevent burglary and robbery . . . $1.00.

CP 3 A SMALL BUSINESS GUIDE TO COMPUTER SECURITY
Addresses issues important to small business owners and recommends ways of deterring computer crime. A checklist on establishing computer security policies is included . . . $1.00.

MARKETING

MT 1 CREATIVE SELLING: THE COMPETITIVE EDGE
Explains how to use creative selling techniques to increase profits . . . 50 cents.

MT 2 MARKETING FOR SMALL BUSINESS: AN OVERVIEW
Provides an overview of the "Marketing" concept and contains an extensive bibliography of sources covering the subject of marketing . . . $1.00.

MT 3 IS THE INDEPENDENT SALES AGENT FOR YOU?
Provides guidelines that help the owner/manager of a small company determine whether or not a sales agent is needed and pointers on how to choose one . . . 50 cents.

MT 6 ADVERTISING MEDIA DECISIONS
Discover how to effectively target your product or service to the proper market. This publication also discusses the different advertising media and how to select and use the best media vehicle for your business . . . $1.00.

MT 7 PLAN YOUR ADVERTISING BUDGET
Describes some simple methods for establishing an advertising budget and suggests ways of changing budget amounts to get the effect you want . . . 50 cents.

MT 8 RESEARCH YOUR MARKET
Learn what market research is and how you can benefit from it. Introduces inexpensive techniques that small business owners can apply to gather facts about their existing customer base and how to expand it . . . $1.00.

MT 9 SELLING BY MAIL ORDER
Provides basic information on how to run a successful mail order business. Includes information on product selection, printing, testing and writing effective advertisements . . . $1.00.

MT 10 MARKET OVERSEAS WITH U.S. GOVERNMENT HELP
Entering the overseas marketplace offers exciting opportunities to increase company sales and profits. Learn about the programs available to help small businesses break into the world of exporting . . . $1.00.

PERSONNEL MANAGEMENT

PM 1 CHECKLIST FOR DEVELOPING A TRAINING PROGRAM
Describes a step-by-step process of setting up an effective employee training program . . . 50 cents.

PM 2 EMPLOYEES: HOW TO FIND AND PAY THEM
A business is only as good as the people in it. Learn how to find and hire the right employees . . . $1.00.

PM 3 MANAGING EMPLOYEE BENEFITS
Describes employee benefits as one part of the total compensation package and discusses proper management of benefits . . . $1.00.

NEW PRODUCTS/ IDEAS/INVENTIONS

PI 1 CAN YOU MAKE MONEY WITH YOUR IDEA OR INVENTION?

This publication is a step-by-step guide which shows how you can make money by turning your creative ideas into marketable products. It is a resource for entrepreneurs attempting to establish themselves in the marketplace . . . 50 cents.

PI 2 INTRODUCTION TO PATENTS

Offers some basic facts about patents to help clarify your rights. It discusses the relationships among a business, an inventor and the Patent and Trademark Office to ensure protection of your product and to avoid or win infringement suits . . . 50 cents.

XIII. INVENTOR'S CLUBS AND ORGANIZATIONS

Many inventor's clubs meet on an informal and irregular basis. There are numerous small groups throughout the United States which fit this category. Check with local schools, in the telephone directory, or with your state economic development office to see if other clubs are currently operating in your city. The larger and more durable organizations are listed here, and many have been active in assisting in the formation of local chapters.

California

Inventors of California
National Congress of Inventors
Organization
P.O. Box 158
Rheem Valley, CA 94570
Norman Parrish
(415) 376-7541

Inventor's Workshop Int. (IWI)
3201 Corte Malpaso
Suite 304-A
Camarillo, CA 93010
Alan Tratner
(805) 484-9786

California Inventor's Council
P.O. Box 2036
Sunnyvale, CA 94087
Barrett Johnson
(408) 732-4314

Colorado

Affiliated Inventors Foundation
501 Iowa Avenue
Colorado Springs, CO 80908
John Farady
(303) 792-7540

Rocky Mountain Inventors
Congress
P.O. Box 4365
Denver, CO 80204
Ken Richardson
(303) 231-7724

District of Columbia

Intellectual Property Owners
1255 23rd St. NW, Suite 850
Washington, DC 20037
Herbert Wamsley
(202) 466-2396

Florida

Central Florida Inventors Council
P.O. Box 13416
Orlando, FL 32859
David E. Flinchbaugh
(305) 859-4855

Tampa Bay Inventors Council
P.O. Box 2254
Largo, FL 34294-2254
F. MacNeil MacKay
(813) 933-9124

Palm Beach Society of American
Inventors
P.O. Box 26
Palm Beach, FL 33480
Kiki Shapero
(304) 655-0536

Georgia

Inventors Clubs of America
P.O. Box 450261
Northlake Branch
Atlanta, GA 30345
Alexander Mannaccio
(404) 938-5089

Hawaii

Inventor's Council of Hawaii
P.O. Box 27844
Honolulu, HI 96827
George Lee
(808) 595-4296

Illinois

Inventor's Council
53 West Jackson, Suite 1041
Chicago, IL 60604-3701

Indiana

The Inventors & Entrepreneurs
Society of Indiana
Box 2224
Hammond, IN 46323
Daniel Yovick
(219) 989-2354

Inventors Association of Indiana
612 Ironwood Drive
Plainfield, IN 46168
Randall Redelman
(317) 745-5597

Massachusetts

Inventors Assoc. of New England
P.O. Box 335
Lexington, MA 02173
Don Job
(617) 862-5008

Worchester Area Inventors
132 Sterling St.
W. Boylston, MA 01583
Barbara Wyatt
(617) 835-6435

Michigan

Inventors Council of Michigan
(INCOM)
2200 Bonisteel Blvd.
Ann Arbor, MI 48109
J. Downs Herold
(313) 764-5260

American Association of Inventors
6562 E. Curtis Road
Bridgeport, MI 48722
Dennis Martin
(517) 799-8208

Minnesota

Minnesota Inventors Congress
Box 71
Redwood Falls, MN 56283
Penny Becker
(507) 637-2344

Mississippi

Inventor's Workshop
P.O. Box 1268
Gulfport, MS 39502
William Baker
(601) 865-0010

Inventor's Workshop
1021 Cedar Hill Drive
Jackson, MS 39206
Tom Howe
(601) 362-2968

Inventor's Workshop
P.O. Box 4398
Meridian, MS 39304
Ed Walters
(601) 483-8241 Ext. 615

Society of Mississippi Inventors
P.O. Box 5111
Jackson, MS 39296
Rick Rommerdale

Missouri

Inventors Assoc. of St. Louis
P.O. Box 16544
St. Louis, MO 63105
Roberta Toole
(314) 534-2677

New Mexico

Albuquerque Invention Club
P.O. Box 30062
Albuquerque, NM 87190
Albert Goodman
(505) 266-3541

Nevada

High Technology Entrepreneurs
Council—(HITEC)
P.O. Box 72791
Las Vegas, NV 89170
George Sanders
(702) 736-3794

Inventor's Workshop
P.O. Box 5531
Incline Village, NV 89450
Charles Earley
(702) 831-2367

New York

Inventor's Workshop
126 Grandview Terrace
Batavia, NY 14020
Robert Bassett
(716) 343-1946

Inventor's Workshop
RFD 2, Box 127
Granite Springs, NY 10527
Chris Novell
(914) 248-5362

Inventor's Workshop
205 S. Central Avenue
Mineola, NY 13116
Ray DiPietro
(315) 656-9210

Ohio

Inventors Clubs of Greater
Cincinnati
18 Gambier Circle
Cincinnati, OH 45218
William M. Selenke
(513) 825-1222 or 922-9462

Inventors Council of Dayton
P.O. Box 77
Dayton, OH 45409
Ron Versic
(513) 439-4497

Inventors Council of Greater
Lorain County
Ohio Technology Transfer Org.
246 Harvard Avenue
Elyria, OH 44035
Dana N. Clarke Sr.

Oklahoma

Oklahoma Inventors Congress
P.O. Box 54625
Oklahoma City, OK 73152
Albert Janco
(405) 848-1991

Invention Development Society
8502 SW 8th St.
Oklahoma City, OK 73128
William Enter, Sr.
(405) 376-2362

Oregon

Inventor's Workshop
20133 NW Morgan Road
Portland, OR 97231
Sig Jensen
(503) 621-3585

Inventor's Workshop
780 SW 231st Avenue
Hillsboro, OR 97123
Larry Campbell
(503) 642-4122

Inventor's Workshop
2876 Cloverdale Road
Turner, OR 97392
Dave Hopfer
(503) 743-2002

Pennsylvania

American Society of Inventors
P.O. Box 58426
Philadelphia, PA 19102-8426
Henry Skillman
(215) 546-6601

South Dakota

South Dakota Inventors Assoc.
P.O. Box 1113
Watertown, SD 57201
Barry Wilfahrt
(605) 886-5814

Tennessee

Tennessee Inventors Association
P.O. Box 11225
Knoxville, TN 37939
Martin Skinner
(515) 584-0105

Texas

Toy & Game Inventors of
America
5813 McCart Avenue
Fort Worth, TX 76133
Bruce Davis
(817) 292-9021

Washington

Inventor's Workshop
2702 40th E.
Tacoma, WA 98404
Ed Rosage
(206) 922-9032

Index

ABOUT THE AUTHOR Robert Park has spent most of his adult life as an inventor, new product developer, prototyper, sculptor and writer.

He has developed numerous new consumer and industrial products including "Col-R-Corn" popping corn, Instant Wild Rice, a sintered bronze tooling process, an artificial "soil" for hydroponic plant growth, a medicinal bacteriostat, a filtered window vent, the 1969 national draft lottery capsules, and several other devices used by military and space agencies.

Park presently works as a consultant and a contract writer.